Advances in
GEOPHYSICS

VOLUME 51

Advances in Geophysics

Volume 51

Series Editor

RENATA DMOWSKA

School of Engineering and Applied Sciences
Harvard University
Cambridge, Massachusetts, USA

AMSTERDAM • BOSTON • HEIDELBERG • LONDON
NEW YORK • OXFORD • PARIS • SAN DIEGO
SAN FRANCISCO • SINGAPORE • SYDNEY • TOKYO
Academic Press is an imprint of Elsevier

Academic Press is an imprint of Elsevier
525 B Street, Suite 1900, San Diego, CA 92101-4495, USA
30 Corporate Drive, Suite 400, Burlington, MA 01803, USA
32 Jamestown Road, London NW1 7BY, UK
Radarweg 29, PO Box 211, 1000 AE Amsterdam, The Netherlands

First edition 2009

Copyright © 2009 Elsevier Inc. All rights reserved

No part of this publication may be reproduced, stored in a retrieval system or transmitted in any form or by any means electronic, mechanical, photocopying, recording or otherwise without the prior written permission of the publisher

Permissions may be sought directly from Elsevier's Science & Technology Rights Department in Oxford, UK: phone (+44) (0) 1865 843830; fax (+44) (0) 1865 853333; email: permissions@elsevier.com. Alternatively you can submit your request online by visiting the Elsevier web site at http://elsevier.com/locate/permissions, and selecting Obtaining permission to use Elsevier material

Notice
No responsibility is assumed by the publisher for any injury and/or damage to persons or property as a matter of products liability, negligence or otherwise, or from any use or operation of any methods, products, instructions or ideas contained in the material herein. Because of rapid advances in the medical sciences, in particular, independent verification of diagnoses and drug dosages should be made

ISBN: 978-0-12-374911-6

ISSN: 0065-2687

For information on all Academic Press publications
visit our website at elsevierdirect.com

Printed and bound in USA

09 10 11 12   10 9 8 7 6 5 4 3 2 1

Working together to grow
libraries in developing countries

www.elsevier.com | www.bookaid.org | www.sabre.org

ELSEVIER   BOOK AID International   Sabre Foundation

CONTENTS

CONTRIBUTORS vii

Chapter 1

Seismicity Induced by Mining: Recent Research

SLAWOMIR J. GIBOWICZ

1. Introduction 1
2. Seismic Monitoring 3
3. Mining Operations and Seismicity 15
4. Source Mechanisms 22
5. Source Time Function 27
6. Source Parameters and Their Scaling Relations 32
7. Precursory Phenomena and Prediction of Large Seismic Events 40
8. Summary 43
   Acknowledgments 45
   References 45

Chapter 2

Induced Seismicity in Hydrocarbon Fields

JENNY SUCKALE

1. Introduction 55
2. Case Studies, Common Observations and Regional Occurrence 57
   2.1. Documented Case Studies 57
   2.2. Common Observations 57
   2.3. Regional Occurrence 67
3. Seismic Monitoring 69
4. Seismicity Induced by Fluid Injection 73
   4.1. Controlled Experiments of Fluid Injection in Hydrocarbon Fields 73
   4.2. Moderate Seismicity Related to Fluid Injection 75
   4.3. Stress Corrosion and Geochemical Processes 77
5. Seismicity Induced by Fluid Extraction 78
   5.1. Fluid Extraction from an Infinitely Extended Horizontal Reservoir 78
   5.2. Fluid Extraction from Axisymmetric Reservoirs 80
   5.3. Fluid Extraction in the Vicinity of Pre-existing Faults 85
   5.4. Analog Models of Faulting Induced by Fluid Extraction 87
6. The Controversy Surrounding Major Midcrustal Earthquakes 87
   6.1. Earthquakes in the Vicinity of Coalinga 89
   6.2. The Gazli Earthquake Sequence 91
7. Conclusions and Outlook 94
   Acknowledgments 100
   References 100

Chapter 3

Phenomenology of Tsunamis: Statistical Properties from Generation to Runup

ERIC L. GEIST

1. Introduction 107
2. General Description of Tsunami Generation Physics 109
   - 2.1. Earthquake Tsunami Generation 111
   - 2.2. Landslide Tsunami Generation 115
   - 2.3. Tsunami Propagation 117
3. Near-field Regimes: Broadside and Oblique 120
   - 3.1. Source Observations—Earthquakes 120
   - 3.2. Source Observations—Landslides 127
   - 3.3. Tsunami Observations—Near-field Broadside Regime 128
   - 3.4. Tsunami Observations—Near-field Oblique Regime 136
4. Far-field Regime 138
   - 4.1. Open-ocean Propagation 140
   - 4.2. Far-field Coastal Interaction 150
5. Summary and Discussion 156
   - 5.1. Observations and Hypotheses 156
   - 5.2. Statistical Descriptions 157
   - 5.3. Knowledge Gaps 158
   - Acknowledgments 158
   - References 159

INDEX 171

CONTRIBUTORS

Numbers in parentheses indicate the pages on which the authors' contributions begin.

GEIST, E.L. (107) U.S. Geological Survey, 345 Middlefield Rd., MS 999, Menlo Park, CA 94025, USA

GIBOWICZ, S.J. (1) Institute of Geophysics, Polish Academy of Sciences, Ul. Ks. Janusza 64, 01-452 Warsaw, Poland

SUCKALE, J. (55) 54-811 Department of Earth, Atmospheric, and Planetary Sciences, Massachusetts Institute of Technology, 77 Massachusetts Ave., Cambridge, MA 02139-4307, USA

SEISMICITY INDUCED BY MINING: RECENT RESEARCH

Slawomir J. Gibowicz

Abstract

The present review of seismicity induced by mining is a continuation of the previous reviews, published in 1990 and 2001 in *Advances in Geophysics*, describing the problems involved and the state-of-the-art of relevant research in this field at the end of 1980s and 1990s. During the last decade, seismic monitoring has been expanded in several mining districts, a number of new techniques have been introduced, and new significant results have been obtained in studies of seismic events induced by mining. This review is organized similarly to some extent to the previous ones. New techniques in seismic monitoring in mines, mining factors affecting seismicity, source mechanisms and source time functions, source parameters and their scaling relations are briefly discussed. Precursory phenomena observed in mines and some attempts at prediction of larger events are reviewed. The new results obtained so far by the Japanese research group in South African gold mines, the concepts of stress diffusion and of "critical earthquakes" applied to seismicity in mines, and numerical modeling of rock mass response to mining are also briefly discussed.

Key Words: Induced seismicity, Seismicity induced by mining, Seismic monitoring in mines, Seismic source mechanisms, Source time function, Source parameters, Seismic scaling relation in mines, Precursory phenomena in mines.   © 2009 Elsevier Inc.

1. Introduction

In the previous reviews of seismicity induced by mining (Gibowicz, 1990; Gibowicz and Lasocki, 2001), the general description of the problems involved and the state-of-the-art of relevant research in this field at the end of 1980s and 1990s, respectively, was given. During the last eight years, seismic monitoring has been expanded in several mining districts, especially in Australia, Russia and China, a number of new techniques have been introduced, and new significant results have been obtained in studies of seismic events induced by mining. I was invited therefore once again by Dr. Renata Dmowska, coeditor of *Advances in Geophysics*, to contribute a new review, describing in some detail the latest achievements in the field.

The first International Symposium on Rockbursts and Seismicity in Mines (RaSiM) held in Johannesburg, South Africa, in 1982 was followed by the second Symposium held in Minneapolis, Minnesota, in 1988, the third Symposium held in Kingston, Ontario, in 1993, the fourth Symposium held in Krakow, Poland, in 1997, the fifth Symposium held again in South Africa in 2001, and by the sixth Symposium held in Perth, Western Australia, in 2005. The proceedings of these last two symposia (van Aswegen *et al.*, 2001; Potvin and Hudyma, 2005) contain some 160 technical papers on the subject, written by experts from over 15 countries. The RaSiM symposia helped us to understand some fundamental ideas about the nature and causes of seismic events in mines. Most important is the recognition that seismicity is the response of the rock mass to stress and strain changes induced by mining, implying that the mine design can influence the level of seismicity. A review of the RaSiM meetings' contribution to the understanding and control of mine seismicity has been published by Ortlepp (2005).

Over the last eight years, a book and a special issue of *Pure and Applied Geophysics* devoted entirely to induced seismicity in a very broad sense have been published. The book, *Induced Earthquakes* by Guha, published in 2000 (Guha, 2000), covers a very wide range of man-made and natural phenomena defined by the author as induced earthquakes, ranging from seismicity induced by water reservoirs and fluid injection into deep wells to underground nuclear explosions and tidal stresses. Seismicity induced by mining is described in a separate chapter that contains interesting information, especially that related to seismicity in the Kolar gold fields of India. The special issue of *Pure and Applied Geophysics*, edited by Trifu, was published in 2002 (Trifu, 2002) and reprinted as a separate topic volume. This volume also covers a very wide range of topics, from mining-induced seismicity to geotechnical applications, from the monitoring of petroleum reservoirs to fluid injections in geothermal areas, and to seismicity associated with water reservoirs. Seven contributions are related to seismicity induced by mining or associated with the presence of mines. A third book, *Seismogenic Process Monitoring*, which is partly devoted to induced seismicity and covers several topics related to seismicity induced by mining, was published in 2002 (Ogasawara *et al.*, 2002a). The book was edited by Ogasawara, Yanagidani and Ando, and presents the outcome of a Joint Japan-Poland Symposium on Mining and Experimental Seismology held in Kyoto in November 1999. It contains 29 papers divided into four parts. The first part, called the near-source monitoring, is composed of 14 contributions, all but one related to seismicity induced by mining in Poland (8 papers) and in South Africa (5 papers).

Two large seismic research projects have been carried out on an unprecedented scale for several years in gold mines in South Africa. Firstly, from 1993 the research on rockbursts and seismicity there is coordinated by the Safety in Mines Research Advisory Committee (SIMRAC). The scope of 94 projects supported by SIMRAC up to 2000 covers assessment of seismic issues, fundamental research into seismic prediction and possible prevention of large events, determination of criteria for mine design to reduce seismicity and rockburst damage, transfer of new technology into the mining industry, and others. Before 1993, rockbursts were either the single greatest or the second-greatest contributor to the fatality rate from accidents in gold mine, while between 1994 and 2000 the fatality rate decreased to half, though it is acknowledged that the SIMRAC research could be just one of the many factors that have contributed to the fatality decline (Adams and van der Heever, 2001).

Secondly, the Japanese research group has initiated semi-controlled seismogenic experiments in South African deep gold mines, where mining takes place at depths of 2000-3600 m, inducing seismic events in the close vicinity of stopes. Thus seismogenic processes can be monitored at very short distances with sensors installed in seismogenic areas. The first experimental field without dykes and faults was established in 1996 at a depth of 2700 m at a mine near Carletonville, in cooperation with the ISS International, where the pilot, first and second experimental phases were carried out (Ogasawara *et al.*, 2001). They have monitored more than 20,000 seismic events with borehole triaxial accelerometers, using a data acquisition system with 15 kHz sampling and a 120 dB dynamic range. Later several experimental fields was established in other mines, one of them near a strong dyke at a depth of 1700 m, and in 2000 the Japanese research group began continuous monitoring of normal and shear strains on a fault, where larger events were expected (Ogasawara *et al.*, 2002b).

This review is organized similarly to some extent to the previous ones. New techniques in seismic monitoring in mines, mining factors affecting seismicity, source mechanisms and source time functions, source parameters and their scaling relations are briefly discussed. Statistical methods and seismic hazard assessment in mines are only touched upon; they were described in some detail in the previous review (Gibowicz and Lasocki, 2001). The other topic not discussed here is the seismic discrimination between underground explosions and seismic events originating in deep mines. Large seismic events in mines are of interest to seismologists monitoring compliance with the Comprehensive Test Ban Treaty (CTBT) prohibiting nuclear explosion testing. There are a few new results related to this topic that have been recently published (see, e.g. Baumgardt and Leith, 2001; Bowers and Walter, 2002; Koch, 2002; Roth and Bungum, 2003; Goforth *et al.*, 2006).

Precursory phenomena observed in mines and some attempts at prediction of larger events are briefly reviewed. The new results obtained so far by the Japanese research group in South African gold mines, the concepts of stress diffusion and of "critical earthquakes" applied to seismicity in mines and numerical modeling of rock mass response to mining are briefly discussed.

2. Seismic Monitoring

Seismicity induced by mining is commonly described as the occurrence of earthquakes caused by rock failures, a result of stress changes in the rock mass near mining excavations. Mining-induced events are caused by increases in the shear stress or decreases in the normal stress acting on the fault planes. Seismic events therefore are induced only in those regions where the ambient stress has been modified substantially by the mine excavations. It is well known that seismic monitoring can provide direct observations related to the occurrence of local instabilities in the rock mass, but a better understanding of the mechanism of induced seismicity is needed to achieve the desired level of control. Deep mines provide unique natural laboratories for studies of seismic source processes and fault properties. The seismic networks there are often three-dimensional, and the observations are made in the source area, allowing a detailed analysis of seismic rupture behavior at the focal depth. Insight into seismicity induced by mining has been expanded mainly through the development of high frequency data acquisition instrumentation with increased dynamic range, and the adaptation of techniques used in analyses of natural earthquakes. This has made possible to record and process small seismic events in mines, with magnitudes less than zero (e.g. Urbancic and Trifu, 2000). As a result, a better understanding of the interaction between excavations, the local stress field, and the local geological structure has been achieved.

During the last decade, seismic monitoring has been expanded in several mining districts, especially in Australia, China and Russia. It is interesting to note that in South Africa, where studies of mining-induced seismicity have been carried out for a long time, some 30 monitoring systems with some 1500 channels provide full waveform records of an estimated 2.5 million seismic events per year (van Aswegen *et al.*, 2001). The quality of waveform data provided by some systems, however, is another matter. The committee commissioned by the Chief Inspector of Mines, South Africa, headed by Durrheim *et al.* from CSIR, South Africa, to lead an extensive investigation into the causes, consequences and risks associated with large seismic events in mining areas in South Africa, such as that of magnitude 5.3 which occurred on 9 March 2005 in the Klerksdorp district, found that "The decline in seismological expertise on mines, at universities, and in research organisations during the past decade is a cause for concern", and made proper recommendations (Durrheim *et al.*, 2006). The Klerksdorp event shook the nearby town of Stilfontein, causing serious damage to several buildings and minor injuries to 58 people. At a nearby deep gold mine, two miners lost their lives and 3200 miners were evacuated under difficult circumstances. The affected mine went into liquidation shortly afterwards, and some 6500 miners lost their jobs (Durrheim *et al.*, 2006).

Potvin and Hudyma (2001) provided an overview of the origin of seismic monitoring in Australia. Although mining seismicity in Australia has been reported for almost a century, until the mid-1990s the problem associated with rockbursts was not perceived to be severe enough to justify investing in seismic monitoring technology. In 1994, the Mount Charlotte Mine near Kalgoorlie was the first mine to install a modern seismic monitoring system in Australia, a system from Miningtek in South Africa. In the same year, an ISS seismic system was installed at Deep Copper Mine to try to understand fault movement near two mine shafts. Installation of ISS systems at Pasminco Broken Hill and North Parkes followed in 1996. During the past years, the use of seismic monitoring has rapidly spread in Western Australia, where a number of narrow-vein gold operations reached depths at which rockburst-prone conditions are present. Mine-wide seismic monitoring systems have now been installed at several Western Australian mines.

In recent years, several nonferrous metal mines passed the depth of 1000 m in China. The hazard related to the ground failures induced by high stresses is at present the most important problem for the deep metallic mining operations in China (Li *et al.*, 2005). An increasing number and severity of rockbursts and rockfalls at the Fankou Lead-Zinc Mine, located in the province of Guangdong, led to the installation of the first multi-channel seismic monitoring system in the Chinese metallic mines. The 16-channel system employs an array of uniaxial accelerometers, fiber optic communication, 20 kHz sampling frequency, and fully automatic event location and source parameter calculation (Li *et al.*, 2005). Rockbursts, often called coal bumps, occur frequently in many underground coal mines in China, but seismic monitoring there has not been carried out. In 2001, an experiment with seismic monitoring was undertaken at the Dong Tan Coal Mine to evaluate its applicability in Chinese coal mining environment. The CSIRO microseismic system was used with four geophones installed in a small area and with 1kHz sampling frequency. The results of this experiment have demonstrated that the applied system can acquire reliable seismic data (Luo *et al.*, 2005).

Intensive mining of the Upper Kama potash deposits in the Perm region in Western Ural, Russia, began in the 1980s, where six mines were in operation (Malovichko, 2005). Ore is extracted from one or two potash beds at a depth between 200 and 400 m below the surface. There was no evidence of any natural seismicity in the vicinity of potash mines prior to mining. The first large felt seismic events induced by mining occurred in the 1990s. The largest event with magnitude $m_b = 4.7$ occurred in January 1995 at Solikamsk-2 mine and was accompanied by a massive roof collapse over a wide area and simultaneous 4.5 m subsidence of the earth's surface above the mine (Malovichko et al., 2001). Since 1995, underground seismic monitoring systems have been installed in all six potash mines. The magnitude-frequency relations there are of bimodal nature. The other characteristics of induced seismicity at potash mines are related to viscous and plastic properties of potassium and rock salt. The seismic activity occurs in areas where mining has been completed some years earlier and not around the stope faces (Malovichko et al., 2005).

In India, the Kolar Gold Fields near Bangalore is one of the deepest and oldest gold mines in the world. Damaging rockbursts with magnitude greater than 5.0 have been generated there and studied extensively for a long time (Guha, 2000). In contrast, at Indian longwall coal mines seismic monitoring systems have been introduced only recently (Sivakumar et al., 2005). A multi-channel 24-bit seismic system was installed in 2001 on a longwall at Rajendra Colliery, situated in Rewa coal fields in central India, to study the rock mass behavior in real time. Seismic data were processed and analyzed to obtain useful information related to prediction of instabilities, location of high stress zones, and fracture propagation in the roof strata.

Various seismic monitoring systems are in operation at different mines and in different countries. From the waveforms recorded by a modern system the origin time, location, radiated energy, scalar seismic moment and other source parameters of a seismic event can be routinely estimated from several associated seismograms, allowing us to contour the relevant parameters, like seismic stress or energy index, on a daily basis.

Tens of seismic arrays are currently in operation in Canada, monitoring rock mass volumes on the order of several hundreds of meters to a couple of kilometers across with a location accuracy of seismic events ranging from meters to a few tens of meters. Seismic systems recording full waveforms are commonly employed. Using 16-bit resolution, 20 kHz sampling, and a combination of uniaxial and triaxial accelerometers, such systems monitor seismic events with moment magnitude between -2 and 0. Larger events are generally monitored by surface geophones. The new generation 24-bit resolution ESG (Engineering Seismology Group Canada Inc.) system offers superior dynamic range, exceptional signal quality, and continuous recording from every sensor in the network. New technological advances in Internet communication have created conditions for continuous seismic data acquisition at sampling frequency of up to 10 kHz (Alexander and Trifu, 2005).

In the United States, an automated PC based seismic monitoring system has been developed by the National Institute for Occupational Safety and Health for use in mine ground control and safety studies. Its structure and its use in underground mines and on the surface have been described by Swanson (2001). At present, the system is clearly outdated and is no longer produced. In the last few years, several ESG seismic systems were installed.

Location of seismic events is the first step in monitoring seismicity in mines and automatic location becomes the only practical solution in mines where the number of events to process and the number of recording channels become too large for manual approach. The accuracy of detection of P- and S-arrival times is critical for automatic location of seismic events. Cichowicz (2005) used several hundred records of seismic events from Bambanani Mine, South Africa, with 27 seismic stations covering an area of 3 by 2 km, to conduct several experiments related to the locations based on first arrivals of seismic waves estimated manually and automatically. He concluded that automatic estimates provide satisfactory results for seismic events recorded by more than six stations. The accuracy of automatic event location is of great practical importance. It is considerably improved by the use of master event relative location and group location techniques. The accuracy achieved by the relative location depends on the accuracy of the location of the master event. An analysis of blast locations showed that both the methods locate seismic events with considerably higher precision than the standard absolute location method (Cichowicz, 2005).

Microseismicity is extensively used to monitor stability of underground structures. Standard location methods cannot be often applied for interpretation of the induced microcracks, since the location accuracy could be on the order of magnitude of the largest observed fractures. Poor knowledge of the velocity of seismic waves is the usual source of the errors involved. Master event location methods offer solution to this problem and significantly improve the locations. Reyes-Montes *et al.* (2005) used a master event location technique to relocate seismic events recorded during the blast excavation of a tunnel at a depth of 420 m at the Underground Research Laboratory in Canada and showed that the resolution was enhanced up to one order of magnitude. The relocated events were located closer to the excavation perimeter than their standard locations, in accordance with the observed damage.

A double-difference (DD) location method was developed by Waldhauser and Ellsworth (2000) for the location of a set of seismic events, when the hypocentral separation between the pair of events is small in comparison to the distance between this pair and the station. All events are located simultaneously by minimizing the double-difference quantity, i.e. the residua between the observed and theoretical travel-time differences for all pairs of considered events. The DD location method reveals much sharper seismic images than the classic joint hypocenter determination technique and is therefore more useful to study geological features and fractures in mines. This technique was used to locate seismic events at several mines in South Africa and Australia (Cichowicz, 2005).

It is well known that the depth of a seismic event is the most difficult parameter for a reliable estimation during the location procedure, especially when an underground seismic network situated at a production level becomes almost planar. The depth is also strongly correlated with seismic velocity and is poorly resolved if the rock mass velocity structure is not sufficiently known (Dębski, 1996). It is interesting to verify whether the DD algorithm can improve the depth accuracy. Rudziński and Dębski (2008) used the DD method to relocate seismic events from Polish coal mines recorded by the regional seismic network of the Central Mining Institute operating in the Upper Silesian Coal Basin. The DD method was enhanced by combining it with the Bayesian approach, considering the a posteriori probability density function (PDF) as the solution of the inverse problem, in order to evaluate the relocation errors (Dębski, 2004). The 2D a posteriori probability density function for the epicenter coordinates of seven events from the Upper Silesian Coal Basin relocated by a DD method combined with the Monte Carlo sampling technique is shown in Fig. 1. The PDF is a measure of the location accuracy. A posteriori PDF for the origin time and depth of an event in the Upper

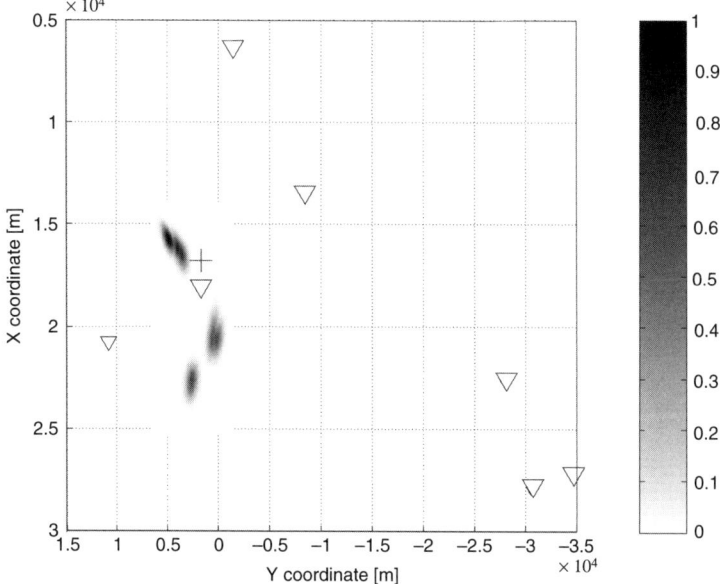

FIG. 1. 2D a posteriori probability density function (PDF) for the epicenter coordinates of seven events in the Upper Silesian Coal Basin, Poland, relocated by a double-difference method combined with the Monte Carlo sampling technique. The PDF is a measure of the location accuracy. Seismic stations are marked by open triangles and the master event is shown by a cross.
Source: Reprinted from Rudziński and Dębski (2008, Fig. 8).

Silesian Coal Basin relocated by a DD method combined with the Monte Carlo sampling technique is shown in Fig. 2. The highest accuracy of the depth determination is observed between 700 and 800 m, the average depth of coal mining in Upper Silesia. There is no correlation between the depth and the origin time in contrast to the classic single-event location methods. In general, the DD method provides more reliable and more stable determination of the depth.

It is well known that mining-induced seismicity follows a multimodal distribution, in contrast to tectonic earthquakes for which the Gutenberg-Richter frequency-magnitude relation has been found to hold for almost all cases, for different magnitude ranges and at various locations. The two major modes of seismicity in mines arise from those seismic events that are associated with geological features and those that are associated with fracturing in the volume of very high stress concentrations ahead of the stope faces. Richardson and Jordan (2001, 2002) showed that events associated with advancing mining at a gold mine in South Africa had moment magnitude smaller than 0.5, called them type A events and interpreted them as "fracture-dominated" rupture events, in contrast to the normal type B "friction-dominated" slip events. These small seismic events were considered by Spottiswoode (2005) as blasts on the evidence that they were the actual development blasts. Partial separation of the two modes can be obtained by selecting a threshold value of magnitude by inspection of the Gutenberg-Richter diagram.

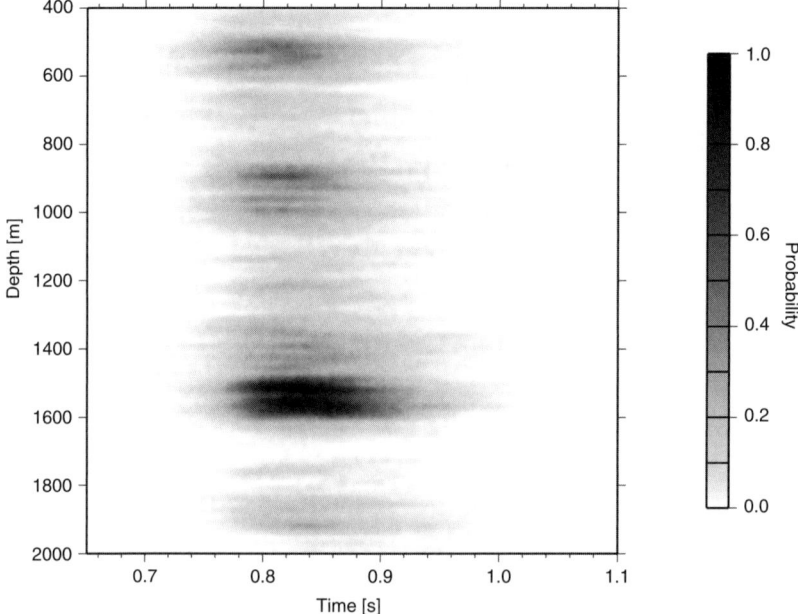

FIG. 2. 2D a posteriori probability density function (PDF) for the origin time T and depth Z of an event in the Upper Silesian Coal Basin, Poland, relocated by a double-difference method combined with the Monte Carlo sampling technique. The PDF is a measure of the location accuracy. There is no correlation between Z and T in contrast to the classic single-event location methods.
Source: Reprinted from Rudziński and Dębski (2008, Fig. 9).

This threshold is usually greater than the threshold of catalog completeness that can be achieved by modern seismic networks in mines. Finnie (1999) has shown that a simple neural network can improve this separation, which is important for statistical seismic hazard analysis and for the identification of active geological structures in mines.

Seismicity in mines is usually highly correlated with mining activity both in space and time. A good example of such correlation is given by Arabasz *et al.* (2005) in a study of seismicity at the Trail Mountain coal mine in Utah. A local seismic network was operated in 2000 and 2001 for continuous monitoring of seismic events associated with longwall mining at a depth of 0.5-0.6 km and for recording high-quality data to develop ground-motion prediction relations. A data set of 1913 seismic events with magnitude smaller than or equal to 2.2 was used to describe the space-time-magnitude distributions of the observed events, which provided a detailed pattern of seismicity associated with a typical longwall coal mining in the area. The epicenter map of 1801 better-located seismic events at the Trail Mountain coal mine, recorded between 3 October 2000 and 30 April 2001, is shown in Fig. 3, reprinted from Arabasz *et al.* (2005). The distribution of magnitude against time for seismicity at the Trail Mountain coal mine from continuous monitoring by the University of Utah regional seismic network, 1999-2001, augmented by the Trail Mountain seismic network, October 2000-August 2001, is shown in Fig. 4, reprinted from Arabasz *et al.* (2005). The correlation between seismicity and mining

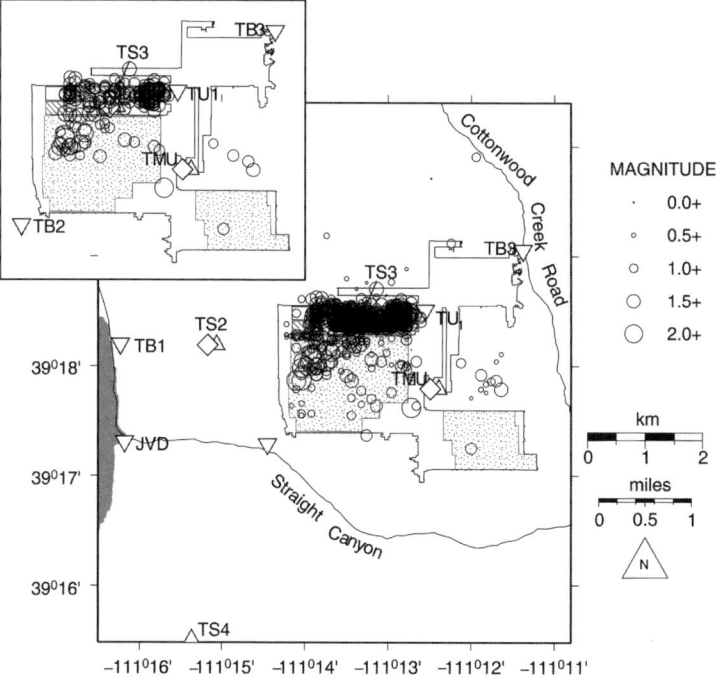

FIG. 3. Epicenter map of 1801 better-located seismic events at the Trail Mountain coal mine, Utah, recorded between 3 October 2000 and 30 April 2001. Inset shows the 216 best-located events. Densest clustering coincides with longwall panels mined during the study period.
© 2005, Seismological Society of America
Source: Reprinted from Arabasz et al. (2005, Fig. 4).

is especially distinct here. Apparent gaps in seismicity during 1999 and 2000 seen in Fig. 4 are inferred to be associated with interruptions in mining or with variations in mine characteristics. The most interesting feature of Fig. 4 in this respect is the reduction in the average upper size of seismicity during the last quarter of 2000 and the first quarter of 2001 in comparison with the preceding quarters. Bold arrows in Fig. 4 mark the period during which the last longwall panel was being mined with a barrier pillar in place. The interruption of mining in the previous panel occurred early in October 2000 and the longwall moved then to the last panel.

Seismicity of the Upper Silesian Coal Basin in Poland provides a good illustration for large-scale regional characteristics of the spatial distribution of seismic events induced by mining. The Upper Silesian Basin, with some 50 underground coal mines in operation, is one of the most seismically active mining areas in the world. Seismic observations date back to the 1950s. Almost 56,000 seismic events with local magnitude greater than 1.5 occurred over the period 1974-2005. Two types of seismic events were typically observed. The so-called "mining-tectonic" seismic events generated by the interaction between mining and tectonic stresses, and "mining" seismic events directly associated with mining works, that occur in the vicinity of mining excavations; their source

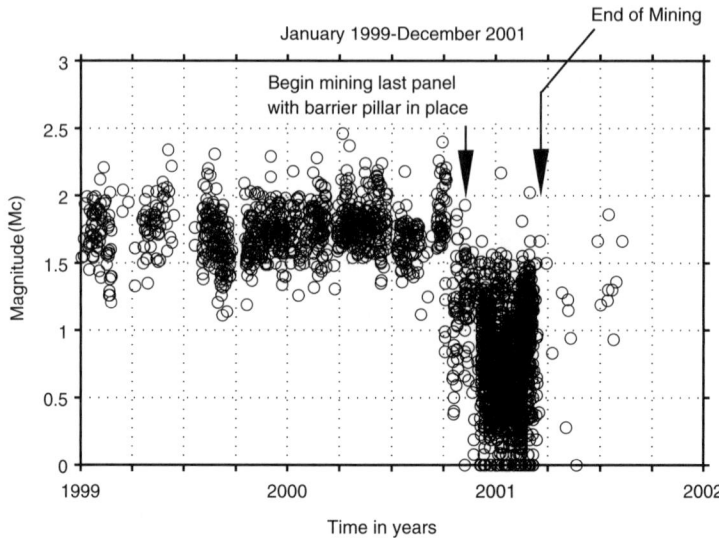

FIG. 4. Magnitude versus time for mining seismicity at the Trail Mountain coal mine, Utah, from continuous monitoring by the University of Utah regional seismic network, 1999-2001, augmented by the Trail Mountain seismic array, October 2000-August 2001.
© 2005, Seismological Society of America
Source: Reprinted from Arabasz et al. (2005, Fig. 7).

mechanisms distinctly differ (e.g. Stec, 2007). There were no reliable historical data related to earthquakes in Upper Silesia available before coal mining was developed on an industrial scale during the second half of the nineteenth century. The spatial distribution of seismic events is not uniform and is of fractal character (e.g. Idziak, 1999; Mortimer, 2002). Despite extensive mining throughout the area, the strong seismic events concentrate in four regions formed by different geological units. The location of these events and the type of their focal mechanism suggest that they are connected with new discontinuities in the rock mass caused by mining (Idziak, 1999).

Seismic events induced by mining are not uniformly distributed not only in space but in time as well. They show the tendency to form nests, swarms, and sequences, similarly to natural earthquakes. There is however a distinct difference between these two types of seismicity. The large events in mines are not always followed by aftershocks in contrast to shallow natural earthquakes. An analysis of over 15 thousands of seismic events recorded at the Rudna copper mine, Poland, between 1980 and 2003, with a threshold magnitude equal to 1.3, shows several types of seismic sequences connected with 345 larger events of magnitude greater than 3.0 (Kwiatek, 2004). Foreshocks were recorded before 58 larger events, usually several tens of minutes before the main events. Aftershocks accompanied 190 larger events and their number clearly depended on the location of the main events within one of 14 seismogenic zones found at the mine, and not on its magnitude. The two largest seismic events of magnitude 4.1 and 4.2 were followed by 9 and 1 aftershocks, respectively. The value of coefficient b of the Gutenberg-Richter relation estimated for different seismic zones ranges from 0.65 to 1.16. The values of coefficient b for seismic events induced in the Ostrava-Karvina Coal Basin, Czech Republic, were estimated by Holub (1999). He found that b values vary in space and time and that they are inversely proportional to the level of observed seismicity, which in turn is influenced by various types of mining activities in the area.

A considerable number of seismic doublets and multiplets within the magnitude range from 0.7 to 3.5 were found at the Wujek and Ziemowit coal mines in Upper Silesia, Poland, and at the Polkowice and Rudna copper mines in the Lubin Mining District (Gibowicz, 2006). They followed the criteria that the difference in magnitude of two events is not larger than 0.15, the distance between their hypocenters not greater than 150 m for coal and 200 m for copper, and the time interval between their occurrence not longer than 10 days for coal and 20 days for copper. The distance and time intervals between two events forming pairs are not dependent on their magnitude. Their focal mechanism is similar in over 60 percent of pairs at coal mines and in about one third of pairs at copper mines. Spatial distributions of doublets display dominant linear trends, characteristic for a given area.

Multiple-event earthquakes are observed not only in natural seismicity but also in seismicity induced by underground mining, although they are usually much smaller. Riemer (2005) examined the records of complex waveforms of seismic events from two different mining areas within the Witwatersrand Basin in South Africa and found the evidence for the presence of two or three subevents in the waveform data. He located each subevent and found that the resulting subevent source areas appear to be associated with different geological and mining situations. The subevents show spatial separation up to 400 m and they may comprise different focal mechanisms. The complex events have important consequences for safety of underground mining operations and for interpretation of related underground damage.

During the last few years, acoustic emission systems monitoring stress-induced microcracking in mines have introduced advanced recording and data processing and analysis methods, similar to those used in monitoring of lower frequency larger seismic events induced by mining. This is a major development in studies of rock mass response to excavation stresses and high-resolution mechanics of microseismic sequences. Unique data from microseismic (MS) and acoustic emission (AE) systems monitoring a common rock volume at the Underground Research Laboratory in Canada, analyzed by Young and Collins (2001); Collins *et al.* (2002); Young *et al.* (2004), have proved to be highly valuable for investigating the temporal fracture mechanics of microcrack formation. The MS and AE systems employed were manufactured by the Engineering Seismology Group (ESG) in Canada. The MS events were found to be spatially associated with clusters of AE events, which occurred up to 6 hours before the MS event and up to 1 hour after, and are termed foreshock and aftershocks, respectively. Each AE cluster was elongated in shape with the long axis between 15 and 50 cm. Figure 5, reprinted from Collins *et al.* (2002), displays a plane view of three clusters showing: (a) the AE events which occurred before the time of occurrence of the MS events, and (b) the events which occurred after the SM events. The dashed ellipses mark the extent of each cluster and the length of the long axis is also marked. The arrows show the direction of movement of the AE event locations. Figure 6, reprinted from Collins *et al.* (2002), shows the cumulative seismic moment of AE events in Cluster 2 in an 8-minute interval surrounding the time of occurrence of the MS events; the AE events are considered as forming foreshock and

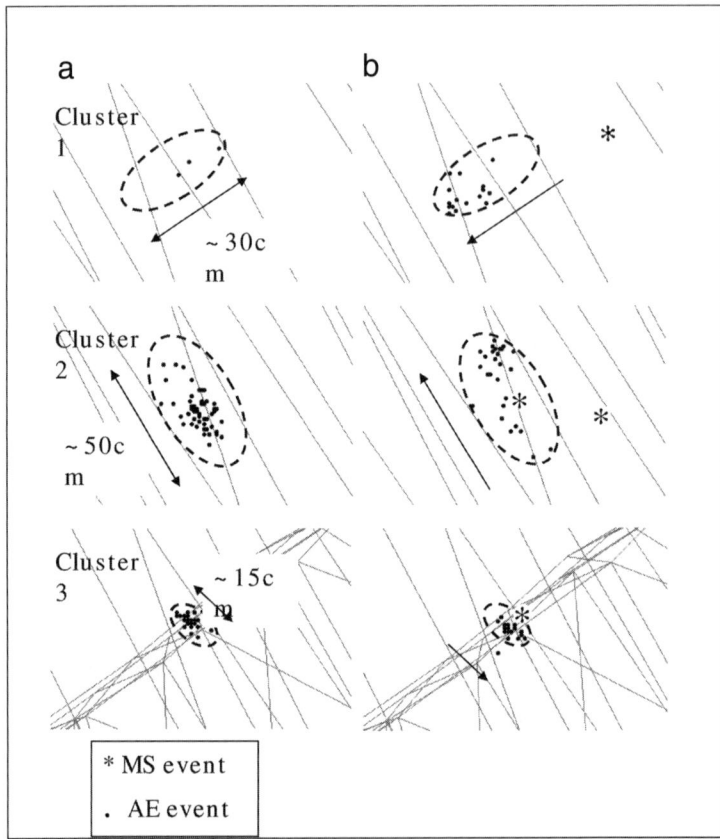

FIG. 5. Plane view of three clusters showing: (a) the acoustic emission (AE) events which occurred before the time of occurrence of the microseismic (MS) events, and (b) the AE events which occurred after the MS events. The dashed ellipses mark the extent of each cluster and the length of the long axis is also marked. The arrows show the direction of movement of the AE event locations.
Source: Reprinted from Collins et al. (2002, Fig. 5) with kind permission from Birkhäuser Verlag AG.

aftershock sequences. Time independent moment tensors were determined for both MS and AE events. The MS events show significant deviatoric sources. The majority of AE events also have deviatoric mechanisms with a few crack opening and crack closure type events. Figure 7, reprinted from Collins *et al.* (2002), shows the stereonet representation of moment tensor solutions of AE and MS events from Cluster 2. The AE events are labeled in order of occurrence. The location of the events is shown in plane view with a star symbol marking the MS events.

Seismic monitoring in mines leads to seismic hazard analysis which is a standard tool used in mines to determine probabilities of occurrence of seismic events that could have an impact on production and safety of miners. More precisely, seismic hazard is the

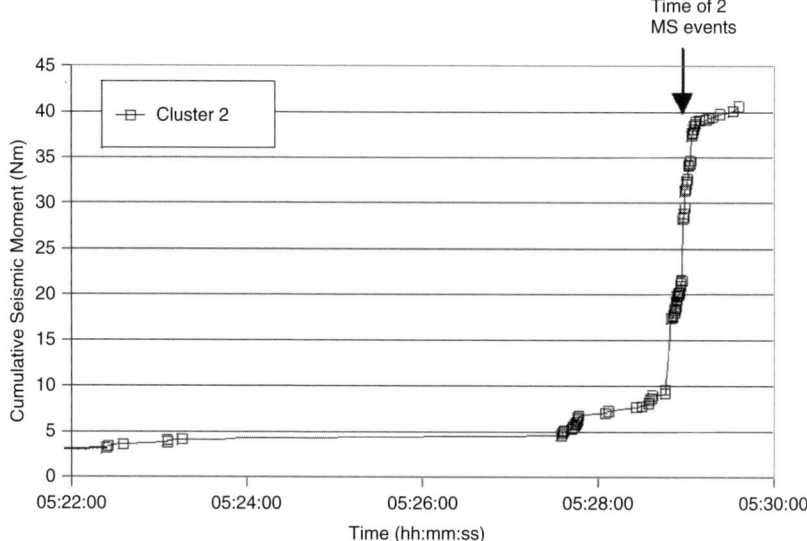

FIG. 6. The cumulative seismic moment of acoustic emission (AE) events in Cluster 2 in an 8-minute interval surrounding the time of occurrence of the microseismic (MS) events. The AE events are considered as forming foreshock and aftershock sequences.
Source: Reprinted from Collins et al. (2002, Fig. 12) with kind permission from Birkhäuser Verlag AG.

probability that a specified value of magnitude of a seismic event will be exceeded during the next specified number of time units (e.g. Kijko et al., 2001), while seismic risk is the product of the probability of occurrence of seismic hazard and the consequences of that hazard (Potvin and Hudyma, 2001). The model of magnitude distribution most widely used in seismic hazard analysis is based on the classic Gutenberg-Richter frequency-magnitude relation, combined with an assumption on the existence of a magnitude upper limit. There is, however, growing evidence that in the case of mining-induced seismicity, the magnitude distribution of seismic events (even of the same A or B type) is often multimodal. A model-free approach to the evaluation of seismic hazard, based on the non-parametric estimator of magnitude distribution was introduced by Kijko et al. (2001) and Lasocki et al. (2002). Both synthetic, Monte Carlo simulated seismic catalogs and real data from a gold mine in South Africa and a copper mine in Poland were used to show that this approach provides seismic hazard parameters with tolerable errors, regardless of whether the magnitude distribution follows the Gutenberg-Richter relation or is multimodal.

Estimates of seismic hazard in many underground mines are based on near-source ground-motion parameters since they are often recorded in close proximity of the actual sources. The input parameters commonly used in such studies are determined from in situ measurements of peak values of ground velocity and acceleration. Talebi and Côté (2005) developed a method for estimating near-source parameters from the records of ground velocity alone, widely available from present monitoring systems in underground

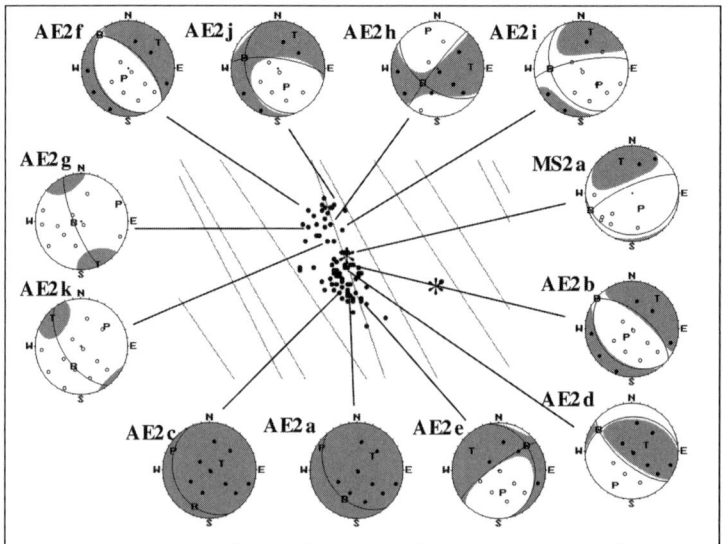

FIG. 7. Stereonet representations of moment tensor solutions of acoustic emission (AE) and microseismic (MS) events from Cluster 2. The AE events are labeled in order of occurrence. The location of the events is shown in plane view with a star symbol marking the MS events.
Source: Reprinted from Collins et al. (2002, Fig. 9) with kind permission from Birkhäuser Verlag AG.

mines. A collection of records of 112 seismic events in the 1-3 magnitude range, recorded at Creighton mine in Sudbury, Canada, was used to measure the source parameters and ground-motion estimates based on this approach. The results are in general compatible with the previous observations of ground-motion parameters.

McGarr and Fletcher (2005) developed ground-motion prediction relations for assessing seismic hazard relevant to seismicity induced by coal mining in the Trail Mountain area in Utah. They used data recorded by a seismic network operated by the University of Utah, which from late 2000 until early 2001 recorded numerous seismic events with magnitude up to 2.2. The ground motion from these events, recorded at hypocentral distances ranging from about 500 m to approximately 10 km, formed the basis for developing new ground-motion prediction relations, augmented additionally by data from a seismic event of magnitude 4.2 in a mine situated about 50 km from Trail Mountain. Using a two-stage regression analysis, McGarr and Fletcher (2005) determined prediction relations for peak acceleration, peak velocity, and pseudo-velocity response spectra, and a peak velocity at a distance of 1 km corresponding to a future earthquake with the probable maximum magnitude.

Rock mass damage assessment is required for many applications in mining engineering practice. While various methods have been used for this purpose, relatively few efforts have been made for using seismic monitoring to quantify the rock mass damage. Cai *et al.* (2001) developed a method of characterizing rock mass damage near excavations based on microseismic monitoring and presented a damage-driven numerical model that takes the seismic data as input to determine damage state described by fracture density. This approach is based on the possibility that a realistic crack size corresponding to a microseismic source is better represented by a tensile cracking model than by the traditional shear model. The rock mass is softened by the introduction of cracks and this is simulated by a micromechanics based constitutive model. The input for the model is the material properties of the intact rock and the location and source size of each observed microseismic event. The model was verified by using data from the Underground Research Laboratory in Canada to investigate the relations between microseismicity, rock mass damage and ground deformation. It was found that predicted rock mass displacements are in good agreement with extensometer measurements.

3. Mining Operations and Seismicity

Mining is not a spontaneous process. Some mining situations may generate more favorable short-term seismic response than others. Mendecki (2005) showed that in one example of tabular sequential grid mining at Mponeng Gold Mine in South Africa, the periods of disorderly blasting of mining panels were associated with lower seismic hazard than when blasting was orderly, irrespective of the rate of mining. He also demonstrated that the seismic data show the presence of long-term correlations, indicating the tendency for the local trend of lower seismic hazard to persist. It may be possible that mining strategies that induce spatial heterogeneity prevent smoothing of the system and are less likely to lead to larger dynamic instabilities.

McGarr (2000) discussed the energy budget of mining-induced seismic events and their interactions with nearby stopes. In situ stress data acquired in the Witwatersrand gold fields, South Africa, indicate that the ambient state of crustal stress there is close to the state of failure in the absence of mining, although the tectonic setting is distinctly stable. Mining initially stabilizes the rock mass by reducing the pore fluid pressure. The extensive mining excavations concentrate the deviatoric stresses in localized regions of the rock mass back into a failure state resulting in seismicity. The energy considerations for seismic events induced by mining indicate that only a few percent is radiated by the seismic waves; its majority is consumed in overcoming fault friction.

Milev and Spottiswoode (2002) considered the effect of rock properties on mining-induced seismicity around the Ventersdorp Contact Reef, Witwatersrand Basin, South Africa. The reef area is characterized by a variety of different rock types that are found above and below the reef in different mining districts. These rock variations are classified into six main geotechnical areas. Knowledge of the difference in seismicity in these geotechnical areas could influence the design of deep-level stopes and optimize the support system. The total seismicity was evaluated using the ratio between the cumulative seismic moment and the volume of convergence, and the seismicity generated in areas with different mining conditions and geology has been compared. Milev and Spottiswoode (2002) paid special attention to evaluate the difference in seismicity associated with different geotechnical classifications. They found that there appeared to be some control of rock types on the considered ratio across two mines that reflected the most variation in rock types. The value of the ratio in areas with soft lava in the hangingwall was about half of its value where only hard lava was present. The ratio was smaller for areas with shale footwall. The data from a larger set of six mines

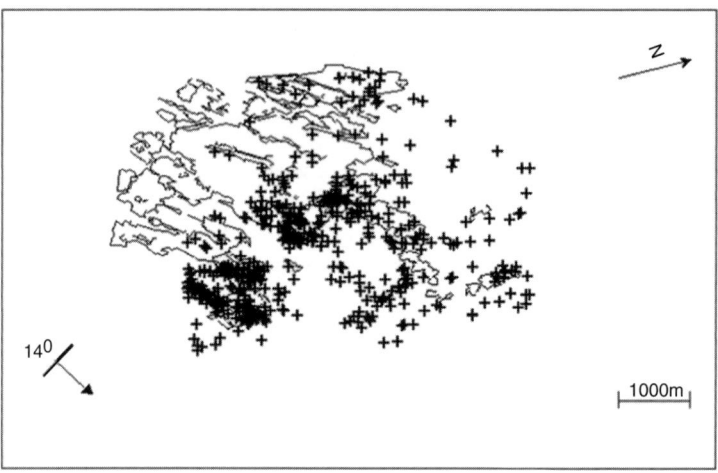

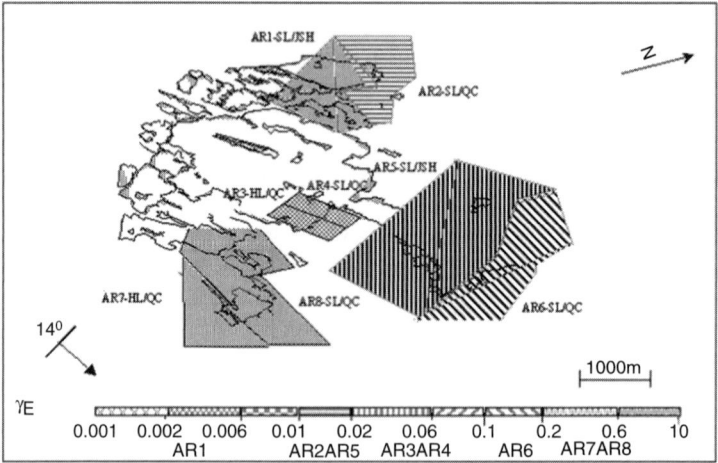

FIG. 8. East Driefontein Mine, South Africa: (a) distribution of the seismic events with local magnitude greater than or equal to 1.0; (b) distribution of parameter γE, the ratio of cumulative seismic moment to the volume of elastic convergence due to mining, for areas corresponding to different geotechnical classifications.
Source: Reprinted from Milev and Spottiswoode (2002, Fig. 1) with kind permission from Birkhäuser Verlag AG.

show that the ratio was lower in areas of lower span, and different types of quartzite footwall did not appear to influence the ratio of cumulative seismic moment to the volume of convergence. Figure 8, reprinted from Milev and Spottiswoode (2002), shows the distribution of seismic events at East Driefontein Mine and of the considered ratio for areas corresponding to different geotechnical classifications. Figure 9, reprinted from Milev and Spottiswoode (2002), presents the cumulative frequency of the ratio values for

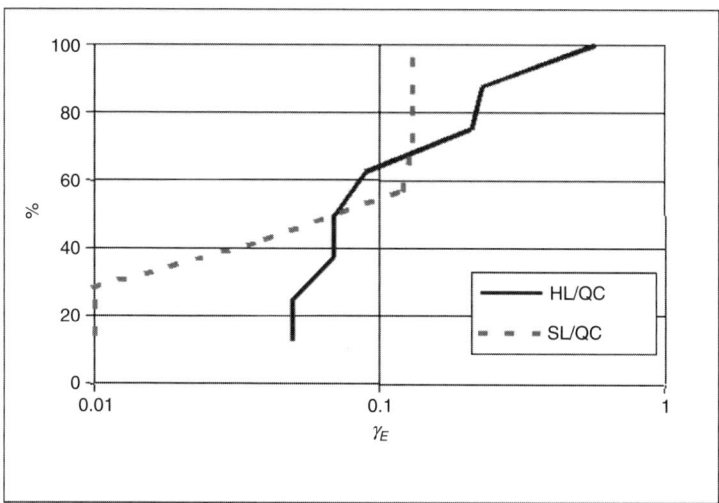

FIG. 9. The cumulative frequency of the γE values calculated for areas with soft and hard lava hangingwall at East Driefontein and Kloof Mines, South Africa.
Source: Reprinted from Milev and Spottiswoode (2002, Fig. 3) with kind permission from Birkhäuser Verlag AG.

areas with soft and hard lava hangingwall at East Driefontein and Kloof Mines. Another parameter, the slope b characterizing frequency-magnitude relations, is presented in Fig. 10, reprinted from Milev and Spottiswoode (2002), where the cumulative frequency curves of the b values are shown, calculated for areas with soft and hard lava hangingwall at East Driefontein and Kloof Mines. It clearly indicates that the b values calculated for areas with hard lava hangingwall are lower than those calculated for areas with soft lava. Higher strength rocks can build up more strain energy before failure, and lower b values imply higher ratio of large to small seismic events.

Durrheim (2001) described the DEEPMINE Collaborative Research Programme launched in 1998 to create the technology and competence required to mine gold safely and profitably at ultradepth (3 to 5 km) in the Witwatersrand Basin in South Africa. The relationship between the depth of mining and the fatality rate has been analyzed in different depth ranges. It was found that there is a strong increase in the rockburst hazard and fatality rate with depth. The research conducted by the DEEPMINE Programme indicates that it will be possible to adapt rock engineering technologies to meet the levels of seismicity expected at ultradepth. Seismic management strategies should be based on prevention and protection rather than prediction. The prevention of seismic events by identifying seismogenic structures (dykes and faults) ahead of mining and the protection of workers and infrastructure using support systems to create rockburst-resistant excavations are the most important strategies for managing the levels of expected seismicity.

As the use of numerical modeling of the behavior of the rock mass around mines or the recording and analysis of seismicity provides on its own only incomplete and limited view on the likely response of the rock for future mining, a number of studies have advocated

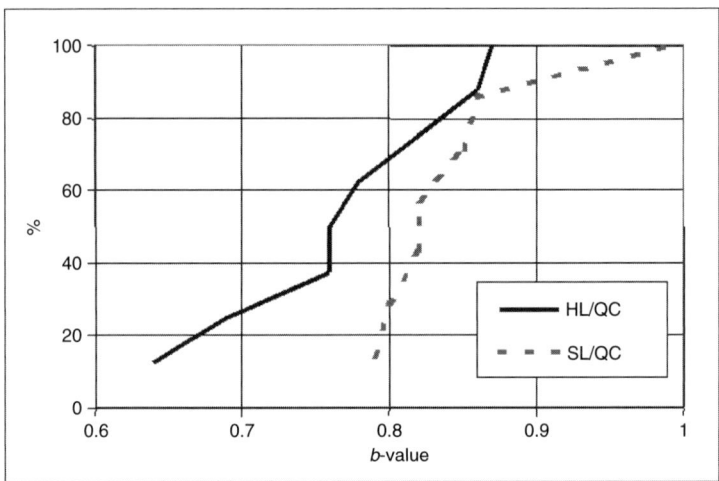

FIG. 10. The cumulative frequency curves of the b values, the parameter characterizing frequency-magnitude relations, calculated for areas with soft and hard lava hangingwall at East Driefontein and Kloof Mines, South Africa.
Source: Reprinted from Milev and Spottiswoode (2002, Fig. 4) with kind permission from Birkhäuser Verlag AG.

moving to an integrated approach using both techniques. Each discipline must be adapted to find common ground and several methods have been proposed in this respect. Spottiswoode (2001) showed synthetic seismic catalogs that mimic observed catalogs in several waves, such as spatial distribution, cumulative seismic moment as a function of mined area, frequency-magnitude distributions and seismicity rate following the blast. The good results from the cumulative seismic moment are particularly interesting for considering mine planning options.

Mine design is based on the assumption that the rock response to mining can be predicted with some degree of confidence and is aimed at reducing the overall seismicity using various criteria. The design criteria are all based on numerical modeling and are sensitive to small changes in rock mass properties and stresses. Seismic data provide a more direct measure of the response of the rock to mining. The use of seismicity data to estimate rock strength was studied by Côté *et al.* (2001) and Wiles *et al.* (2001). Lachenicht *et al.* (2001) presented a quantitative analysis of seismicity resulting from fault slip. Hofmann *et al.* (2001) compared the spatial and temporal distribution of observed seismic events to the amount expected from stress and energy changes within the rock mass. Senfaute *et al.* (2001) described detection and monitoring of high stress concentration zones at Lorraine, France, coal mines using numerical and seismic methods.

Spottiswoode (2005) described the principles of methodology for an integrated analysis of both rock mass modeling and observed seismicity to predict the amount of seismicity that will be likely to occur over the subsequent months in a given area. Mercer and Bawden (2005a,b) developed a statistical approach for the integrated analysis of mining-induced seismicity and numerical stress estimates using data from

Creighton Mine in Sudbury, Ontario. The overall objective of their study was to develop a methodology for identification and evaluation of meaningful relations between variables from the seismic and stress data. Specifically, the model adequacy and stability of each relation were evaluated. All but three out of eleven of the developed relations met the statistical requirements for relations of this type and all were shown to be quite stable.

Various strategies are designed to reduce the occurrence of damaging seismic events and their effects at different mines. At Mponeng Mine, producing gold from the Ventersdorp Contact Reef, South Africa, the adopted dip-pillar based layout has been highly successful. A dramatic decrease in the number of damaging events and a reduction in seismic-related injuries have been achieved (McGill, 2005). In the Klerksdorp area, South Africa, the use of scattered mining was necessitated by the geological complexity of this part of the Witwatersrand Basin. Several analyses were conducted using various statistical and seismicity parameters to describe and compare different mining situations and the rock mass response to production (Dunn, 2005). A rock engineering interpretation of seismic observations was attempted in terms of anticipated stress levels, deformation and relative seismic hazard. It was found that some parameters were good indicators of the local rock mass stiffness and could be related to areas where high levels of stress-induced fracturing were expected. Chen *et al.* (2005) described the experience with mining-induced seismicity at the Darlot Gold Mine, Western Australia, at a relatively early stage of mining operations and at increasing levels of extraction. One of the unique characteristics at Western Australian mines is the influence of high stresses at relatively shallow depths. Planning and sequencing of stopes are used to manage and reduce hazards associated with a rockburst-prone rock mass and large geological structures. Seismic monitoring systems are used for the analysis of seismic data and the identification of potentially hazardous areas. Pytel (2003) presented a numerical model of the rock mass response to mining at Polish copper mines in the Lubin Mining District, where the room-and-pillar technique is generally used for ore extraction. A high number of barrier pillars of different sizes left there creates local stress concentrations and high seismic hazard in the adjacent areas. A 3D numerical model, based on the finite element method, is used to estimate the static and dynamic response of the rock mass to mining operations.

The mechanism of caving, used for ore extraction at block caving mines, could not be explained before the introduction of modern seismic monitoring systems. Duplancic and Brady (2001) analyzed field data collected from the Northparkes Endeavor 26 block cave, Australia, using various instruments including seismic monitoring. The seismic monitoring provided a unique 3D view of the behavior of the rock mass and an insight into the mechanics controlling cave propagation. The analysis of stress and seismic data has shown that different failure mechanisms can be used to describe cave propagation. Slip along pre-existing discontinuities was shown by analysis of seismicity to be the main failure mechanism. Stress analysis showed that this mechanism did not extend far ahead of the propagating cave front. Dunlop and Gaete (2001) described a simple estimator that has been developed at El Teniente Mine, Chile, to evaluate the seismic rate associated with caving mining. The seismic rate was related to the mining intensity, which in turn was related to the factors contributing to the rock mass loss of equilibrium conditions. The rate was used to adjust the middle-term mining plan variables in order to control the rock mass seismic response. Iannacchione *et al.* (2005) analyzed a sequence of seismic events caused by roof falls at the Springfield Pike Quarry, Pennsylvania, an underground room-and-pillar mine. Seismicity associated with roof falls is considerably different from the other types of seismicity induced by mining. The roof fall events are composite, comprising multiple impacts of rocks of varying size on the mine floor. They are very inefficient sources of high frequency seismic radiation and therefore are characterized by very low apparent stress, and their seismic signatures are emergent and of long duration in comparison with shear rupture events.

Active faults are not accessible at the focal depth of natural earthquakes and it is not possible therefore to relate directly the estimated source parameters to the associated rupture. But the source region of large seismic events in deep mines is often accessible. In very deep mines where large areas of massive hard rock are present without significant faulting or dykes, rockbursts are more compact but often more violent in nature than those associated with geological faults Ortlepp (2000). An extensive unique study of a rockburst rupture was carried out in situ at East Rand Propriety Mines, South Africa, from 1974 to 1975, and was described by Gay and Ortlepp (1979) and Ortlepp (1992, 2000). The study involved exploration of the nature and extent of two fresh shear ruptures traversing the rock mass ahead of the faces of a longwall stope in an uncomplicated geological setting at a depth of 2050 m below the surface. The ruptures were essentially planar features of considerable spatial extent with normal dip-slip displacement of up to 0.1 m. This early study remains unique in respect of the extent of exposure of the shear surfaces and the number of detailed photographs taken of its morphology, textures and displacements. Certain striking features were revealed by a scanning electron microscopic study of the fresh cataclastic rock-flour which formed part of the comminuted filling of these ruptures (Ortlepp, 2000).

In August 1998, a major deformation zone was exposed over several meters during mining operations at a depth of 2463 m at Western Deep Levels Gold Mine, South Africa. This product of a recent rockburst was studied by Stewart *et al.* (2001). They found that this zone consisted of three shear zones, with dip-slip displacements of up to 0.15 m, that were oriented near-parallel to the advancing stope face. Jogs and highly pulverized, cataclastic rock-flour were developed on the displacement surfaces, and several sets of secondary extensional fractures occurred on either side of the shear zones. The initial deformation experienced by this zone was brittle and tensile, was related to stresses induced by mining, and has been masked by later changes in the stress field, which resulted in shearing. This deformation zone seems not to be controlled by pre-existing geological features and represented therefore a rockburst fracture, which is believed to be related to a seismic event of magnitude 2.1 recorded in July 1998, though such a slip is highly unlikely to be associated with a so small seismic event (see McGarr and Fletcher, 2003).

To advance understanding of the mechanism of origin of a rockburst, a deliberate search was made for a pristine shear rupture, described by Ortlepp (2001). In September 1998, a significant series of three rock bursts was identified at a depth of 2555 m below the surface at Western Deep Levels Mponeng Mine. Photographic, petrological and mineralogical studies were made of a typical stope-induced shear rupture and its associated gouge material. Most of the diagnostic features on the macroscale were found to be virtually identical to those investigated some 25 years earlier at East Rand Propriety Mines in central Witwatersrand (e.g. Ortlepp, 2000). From further close similarities visible on the microscopic and sub-microscopic scale it follows that the shear rupture had been driven through an intact undamaged rock material in both cases. From this and other evidence, it could confidently be inferred that the origin of the rupture was some distance from the plane of the excavation and that the fracture front propagated towards the stope area.

Dor *et al.* (2001) studied the Dagbreek fault at Matjhabeng Mine, Welkom Gold Field, South Africa, that slipped during the April 1999 Matjhabeng seismic event of magnitude 5.1. The fault zone that was active during this event was up to 30 m wide and it contained tens of gouge zones that are characterized by the well-known "rock-flour" and quartzite fragments. The slip on individual gouge zones ranged from a few millimeters to 21 cm, and the largest total displacement along the Dagbreek fault was 44 cm.

An important type of severe rockburst in deep, tabular mines in South Africa, resulting from the sudden eruption of a dynamic brittle shear zone, was described by Ortlepp *et al.* (2005). Earlier studies involving thin-section micrography and scanning electron microscopy of the fault gouge and slip surfaces there had shown the presence of unique sub-microscopic particles. They indicated extreme processes accompanying the evolution of the fracture. The origin of these particles has important implications for the understanding of the large stress drops and unusually violent damage associated with this particular type of rockburst. Ortlepp *et al.* (2005) showed that the sub-particles are characteristic of induced dynamic brittle shears in quartzite. Using an energy balance analysis, they quantitatively supported the hypothesis that the polyhedral sub-particles are the result of a shock unloading phenomenon. This has important implications for the frictional resistance on the fault surface.

Kersten (2005) studied the geometry of rockburst fractures corresponding to the data obtained from the work of Ortlepp (2000) related to East Rand Propriety Mines, South Africa. Rockburst fractures in a homogeneous continuous quartzite are composed of three distinctly defined fracture sets: continuous fracture surface, en-echelon configuration, and relatively short fracture surfaces with distinct branching. The relative proportion of the various sets and their geometry determines the manner of propagation of a rockburst fracture. Kersten (2005) postulates that the relationship between the orientation of the maximum principal stress and the direction of the various subsets defines the size of the fracture. The orientation of the maximum principal stress and the optimum failure surface have to coincide over a distance of twenty meters to cause a rockburst of the order of magnitude 2. Modeling shows that these conditions are predictable.

A last example of a study in the source area comes from the Underground Research Laboratory in Canada. Collins and Young (2000) analyzed seismicity induced by a tunnel excavation through two lithological units, granite and granodiorite, in a high stress environment, in an attempt to understand lithological differences in the zone of damage. The events in the granite and granodiorite were extremely small, having a similar range of magnitude from about −4 to −3 and the source radius from about 0.1 to 0.5 m. The seismic response differed significantly in the granite and granodiorite, with excavations in the granite having more seismic events occurring ahead of the tunnel face, and a shorter overall temporal response. Petrographic analysis of the rock samples showed stress relief microcracking predominantly in the larger quartz crystals, suggesting that these are the weakest mineral grains. The smaller-grained, more homogeneous granodiorite showed less stress relief microcracking. From the seismic and petrographic evidence, Collins and Young (2000) concluded that the crack initiation stress is lower in the granite than in the granodiorite.

4. Source Mechanisms

Studies of the source mechanism of seismic events in mines play an important role in understanding the various modes of rock failure observed in underground mining environments. The moment tensor inversion technique is the most effective methodology in this respect and has been used extensively since the early 1990s to study the source mechanism of seismic events induced by mining (e.g. McGarr, 1992). Modern hardware and software technology made possible the use of seismic monitoring in a large number of applications related to induced seismicity, where hundreds of events per day are routinely recorded. For such applications, fast and automatic techniques are needed. Trifu *et al.* (2000) proposed an original method for the evaluation of moment tensor components using time-domain calculations of low frequency displacement amplitudes and taking account of body-wave polarities. This approach is reliable in retrieving the geometrical aspects of the seismic source. Inversions for the pure shear mechanism and a general mechanism are shown to be robust for various wave-type solutions.

Dahm *et al.* (1999) investigated two methods for an automatic moment tensor inversion. The first is a single-source inversion using amplitude spectra of body waves and theoretical Green's functions, and the second is a relative moment tensor inversion using spectral moments of body waves. Both methods were successfully tested on synthetic data and were applied to hydraulically induced small seismic events at the Bernburgat salt mine, Germany, at a depth of 400 m. The estimated moment tensors of some 190 events showed similar radiation patterns with a major double-couple component and a small or zero isotropic component.

Andersen and Spottiswoode (2001) introduced a new hybrid moment tensor inversion method based on automatic processing. Firstly, absolute moment tensors were determined using input data consisting of spectral plateaus and first directions of ground motion. Corrections were then made to the input data on median values of the ratios of theoretical to observed values. The corrections were introduced in stages to maintain stability and to track changes in the moment tensor solutions. The aim of this method was to reduce the effects of systematic errors occurring at a recording site on the moment tensor elements. The hybrid moment tensor solutions for a set of 10 events from a cluster at Oryx Gold Mine, South Africa, showed distinct similarity and changes with time that were not apparent for the absolute solutions.

The most general seismic source is a combination of a double couple (sudden displacement on a fault plane), a compensated linear vector dipole (sudden change in shear modulus in the presence of axial strain), and a volumetric source (sudden volume change). Non-double-couple mechanisms of seismic events in mines are indicators of local modifications of the stress field induced by excavations. But non-double-couple components may also arise as a result of inadequate station coverage, source mislocations, the presence of noise in the data and inadequate modeling of rock mass structure. Panza and Sarao (2000) reviewed possible sources of false non-double-couple solutions and estimated their reliability from an error analysis. Their analysis of synthetic data and real observations from Italian volcanic and geothermal areas showed that false non-double-couple components can be recognized in the moment tensor solutions. The analyses of real data must be proceeded by synthetic tests to define the lower limits above which the non-double-components can be considered statistically significant at a given confidence level.

Anisotropy is often present in the mining environments, and is frequently neglected in moment tensor inversions. Šileny and Vavryčuk (2002) found that a neglect of anisotropy of the rock mass can have considerable effect on the source mechanism derived from the waveform inversion. If the source is a pure double couple, spurious non-double-couple components appear. Their magnitude increases with the degree of anisotropy. The orientation of the retrieved double couple is also degraded. To reduce these effects, an asynchronous alignment of the three-component waveforms is proposed in a two-step inversion that firstly, linearly determines the moment tensor rate functions followed by the nonlinear determination of the spatial solution and the source time function.

Šileny et al. (2001) studied the effects of ignoring the reflecting interface in the rock mass model. Such a situation often occurs when seismic events located close to mine excavations are recorded and processed using a simplified model either describing the interface approximately or ignoring it completely. A numerical study was performed, in which the wavefield generated by a source situated close to a free surface on a homogeneous half-space was treated as the observed data. The free surface was then ignored when Green's function was constructed for the point-source inversion. The configuration of the Underground Research Laboratory in Canada was simulated. It was found that gross features of the source, such as the orientation of double couple and the general shape of the source time function, can be retrieved satisfactorily for the hypocenter correctly located. But the formal error analysis yields rather large error estimates as a result of the omission of the free surface, providing acceptably constrained solutions only at about 70 percent confidence level. As a consequence of the free surface neglect, spurious non-double-couple components appeared in the solutions as well.

In general, therefore, the seismic sources retrieved by the waveform inversions may contain errors caused by incomplete knowledge of the medium along the propagation paths of seismic waves, generating in turn distortions to the calculated moment tensors. But on the local scale, the problem of moment tensor resolution seems not to be too serious. Šileny and Milev (2006) investigated the distortion of moment tensor on a local scale by inverting the records of a large explosion at Kopanang Gold Mine, Klerksdorp region, South Africa. The source isotropic component was found to be dominant, whereas the deviatoric components were spurious. An analysis of their stability indicated that they were not significant; uncertainty of 5 percent in seismic velocities and of 10 percent in attenuation within the homogeneous model of rock mass at the mine provided the non-deviatoric mechanism consistent with the blast. The homogeneous model, however, can only be applied to seismic records from close stations, within a few kilometers from the source. The records from distant stations were too complex for modeling by a homogeneous rock mass.

The monitoring of seismicity in mines commonly employs uniaxial sensors along with a limited number of triaxial sensors. For any given number of channels, the use of uniaxial sensors provides a denser network configuration that ensures a higher sensitivity and increased event location accuracy. Trifu and Shumila (2002b) showed that the use of uniaxial recordings maintains the linear dependence of the low frequency displacements, with first polarities included, on the moment tensor components, and can be readily incorporated into the evaluation of failure mechanism using moment tensor inversion. Synthetic data analysis supports the use of a linear propagation approximation for the error evaluation in the orientation of principal stress axes. Case studies from the Kidd

Mine, Ontario, indicate that the general characteristics of the mechanism solution were well retrieved, as relatively small disorientation angles of 15-20 degrees were found between the solutions derived from rotated triaxial data and those from unrotated triaxial data used as uniaxial data, and 25-30 degrees between the solutions based on rotated triaxial data and those on independent uniaxial data.

The classic double-couple seismic source model is not always representative of mining environments. The general dipole source model is more suitable since it contains the isotropic part and the linear dipoles. These components can describe explosions and collapse of cavities. The focal mechanism of some seismic events induced by mining can differ from the dipole model as well. The equivalent single force model can then be used to describe some of these events. Šileny and Milev (2005) proposed a unified methodology to determine the parameters of both dipole and single force models based on the inversion of waveforms generated by seismic events in mines. Preference of the dipole or single force source model is assessed from the stability of the retrieved source orientation from both P and S waveforms, and from P waveforms alone. This approach was applied to the waveforms of seismic events recorded at Driefontein Gold Mine, South Africa, and was found to be highly effective.

Hazzard and Young (2002) presented a numerical modeling approach that simulates cracking and failure in rock and the associated seismicity and described a technique for the evaluation of moment tensors of synthetic seismic events generated by micromechanical modeling. The proposed technique provides a potentially objective validation of the model by comparison with the results derived from real recordings. The test study of a notch formation revealed several implosional events, in partial agreement with recorded seismicity, and showed that these events initiated as tensile cracks but evolved into implosional regimes when a volume compensation occurred.

McGarr (2005) reviewed some essential aspects of the three types of focal mechanisms computed by the moment tensor inversions, and considered how these diverse moment tensors are related to the mining operations and their associated ground deformation. The first type, the pure deviatoric mechanism associated with either the slip across a pre-existing fault or the slip associated with fresh-rock fracture, is thought to be the result of stress perturbations induced by mining but is sufficiently remote from the nearest mine opening and the focal mechanism is controlled mostly by the ambient state of stress. The second type, the mechanism composed of deviatoric and implosive components, involves strong interaction between the seismic event and the adjacent mine stope and involves a combination of fault slip and coseismic stope closure. The third type, the mechanism composed of deviatoric and explosive components, is characteristic of the microseismicity associated with the time-dependent development of a fracture zone in the immediate environment of mining excavations.

Fletcher and McGarr (2005) performed moment tensor inversion of the displacement waveforms of six seismic events from the Trail Mountain coal mine, Utah. Their moment magnitudes ranged from 1.3 to 1.8, and they occurred at a depth of 0.2 to 0.6 km. Four of the six events showed a substantial volume reduction, presumably due to coseismic closure of the adjacent mine openings, ranging from 27 to 55 percent of the shear component. Apparent stresses ranged from about 0.02 to 0.06 MPa, which is at the low end of the values compiled by McGarr (1999). The energy released by each event, approximated as the product of volume reduction and overburden stress, compared with

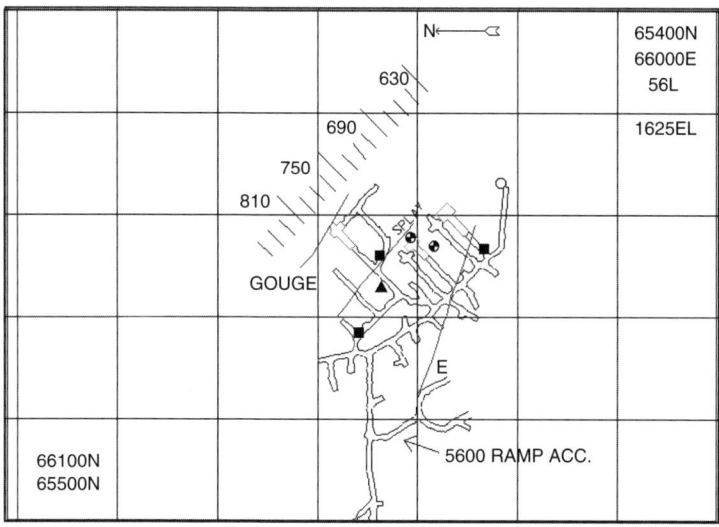

FIG. 11. Plane view of the seismic event locations within close proximity to the 5600 level at Kidd Mine, Ontario. Locations of the sensors are marked by squares and triangles. Two marked events aligned along one of the drifts exhibit pure shear failure source mechanism.
Source: Reprinted from Trifu and Shumila (2002a, Fig. 6) with kind permission from Birkhäuser Verlag AG.

the corresponding seismic energy revealed seismic efficiencies ranging from 0.5 to 7 percent, consistent with the upper bound of 6 percent proposed by McGarr (1999) for crustal earthquakes.

A novel moment tensor inversion approach (Trifu *et al.*, 2000) was applied by Trifu (2001) and Trifu and Shumila (2002a) to small seismic events, with moment magnitude ranging from -1.2 to 0, from the Kidd Deep Copper-Zinc Mine, Ontario, extending to depths over 2000 m. The data consisted of 86 and 35 events, respectively, recorded on 8-10 triaxial accelerometers installed underground. Reliable solutions were obtained for 58 and 21 events, of which 38 and 14 showed pure shear mechanisms and 19 and 7 exhibited a significant positive volumetric component with a small shear component. Six events out of seven with large volumetric components were located within a sill pillar on the 4700 level (1450 m) and had subvertical pressure and subhorizontal tensional axes, in agreement with the presence of tensile cracks close to openings for incipient pillar bursting. The pure shear events were located outside the sill pillar, between the 4600 and 4800 levels (1415-1475 m), and on the 5600 level (1725 m) within a highly fractured rock mass. For these events, the subvertical nodal planes matched closely the orientation of subvertical NW-SE fractures aligned parallel to the major faults in the area. A plane view of the seismic event locations within close proximity to the 5600 level at Kidd Mine is shown in Fig. 11, reprinted from Trifu and Shumila (2002a). Locations of the sensors are marked by squares and triangles. Two events aligned along one of the drifts, marked in Fig. 12, exhibit pure shear failure source mechanism shown in Fig. 12, reprinted from

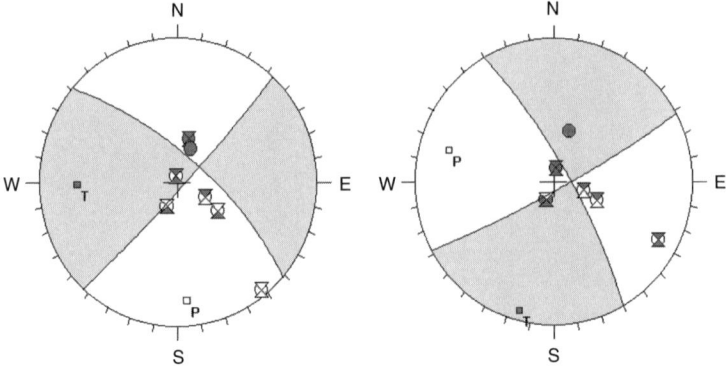

FIG. 12. Source mechanism of the pure shear type for two seismic events at Kidd Mine, Ontario. The nodal planes are oriented approximately parallel and perpendicular to the nearby major faults. *Source:* Reprinted from Trifu and Shumila (2002a, Fig. 7) with kind permission from Birkhäuser Verlag AG.

Trifu and Shumila (2002a). The nodal planes are oriented approximately parallel and perpendicular to the nearby major faults.

Richardson *et al.* (2005) performed moment tensor inversion for 14 seismic events with moment magnitudes between 1.4 and 3.0 from the Far West Rand gold mining region of South Africa. These events occurred at depths between 1 and 4 km and were recorded locally by four networks of 102 three-component geophones installed throughout the mines. The moment tensors of these events were consistent with pure double-couple solutions. No significant isotropic components of the source have been found. This result supports the hypothesis that large mining-induced rock failures are generated by shearing on pre-existing planes of weakness under frictionally controlled conditions. A surprising number of strike-slip mechanisms were determined, instead of normal faulting, which was expected.

In contrast to these results reported from South Africa, there is growing evidence that some seismic events in mines show dilatational first motions of *P* waves at all recording stations. Talebi and Côté (2005) described an investigation of implosional source mechanisms often observed at a hard-rock gold mine in Quebec, Canada, where double-couple events are also generated. The seismicity originated at depths of 1.0-1.3 km and was monitored by a network of 5 triaxial sensors and 15 uniaxial sensors located around the seismically active levels. A data set containing 9 events with double-couple focal mechanisms and 9 events with purely dilatational *P*-wave first motions with magnitude ranging from 0 to 1 was analyzed. Various source models were considered and the ratio of the energy of *P* and *S* waves was used as an indicator. The results implied that the implosional events could be described by a force dipole modeling of a sudden contraction of tabular cavities within the mine.

Sprenke *et al.* (2002) made a comparison of body-wave displacement with damage observations from a rockburst at the Lucky Friday Mine, Coeur d'Alene Mining District, Idaho, where reconciling seismic fault plane solutions and a rockburst damage has been problematic. Much of the confusion arose from the presence of mining-related fractures in the vicinity of openings which form planes of weakness for seismic slip and which allow the stope margins to implode violently into the openings. A shear implosional model with an axis of maximum principal extension oriented normal to bedding, explains the seismic data, the damage to the stope margins, and the observed bedding slip displacement in the access ramp.

An interesting example of complex non-double-couple focal mechanism was studied by Teyssoneyre *et al.* (2002) who carried out moment tensor inversion of regional body waveforms to retrieve the source mechanism of the 1996 seismic event of moment magnitude 4.8, which occurred in the closed Teutschenthal potash mine near Halle, Germany, where an area of 2.7 km^2 collapsed, following the sudden breakage of carnallite pillars at a depth of 700 m. A source time function was obtained, which was decomposed into two subevents. The first subevent had a large positive isotropic component and a deviatoric component with near vertical nodal planes, shown in Fig. 13, reprinted from Teyssoneyre *et al.* (2002). No volume change was found for the second subevent, and its deviatoric mechanism is close to a pure double-couple source, as shown in Fig. 14, reprinted from Teyssoneyre *et al.* (2002). The inversion was based on *P* waves and was stable; the addition of *S* waves did not change the results. The obtained moment tensor is compatible with a large cavity collapse.

Malovichko (2005) investigated the mechanism of so-called "low frequency" seismic events that occur at mines of the Upper Kama potash deposits in Western Ural, Russia. Intensive low frequency (0.5-2 Hz) waves, which were interpreted as Rayleigh waves (Malovichko and Baranov, 2001), are observed on the waveforms of some events routinely recorded by seismic networks. Classic source models in the form of expanding shear cracks do not reproduce the observed waveforms. Other dynamic processes that occur and are recorded at potash mines, such as pillar bursts, rock falls and floor upheavals, were analyzed and equivalent point sources were determined for these processes. A comparison of synthetic waveforms corresponding to these sources with observed waveforms showed that only rock falls, modeled by the source in the form of a vertical single force, could generate intensive Rayleigh waves.

The source mechanisms of seismic events induced by longwall coal mining in the Upper Silesian Coal Basin, Poland, have been intensively studied and was recently described by Stec (2007). Various types of reliable source solutions depend on local tectonics, source location with respect to the geometry of mining operations, past mining geometry, and preventive measures applied in the mine. Large events are often characterized by normal dip-slip mechanism with a marked strike-slip component. The azimuth and dip of their rupture planes show good correlation with the strike and dip of the pre-existing local faults. The other type of seismic event is connected directly with mining operations and occurs in the close vicinity of mining works. These are small events and their source mechanisms often have large explosive components.

5. SOURCE TIME FUNCTION

The Source Time Function (STF) of a seismic event describes the moment rate release and the rupture evolution in the source. If directivity effects are observed, the rupture velocity and rupture direction could be estimated. The extraction of STF from seismograms requires separation of the source effects from those of the path,

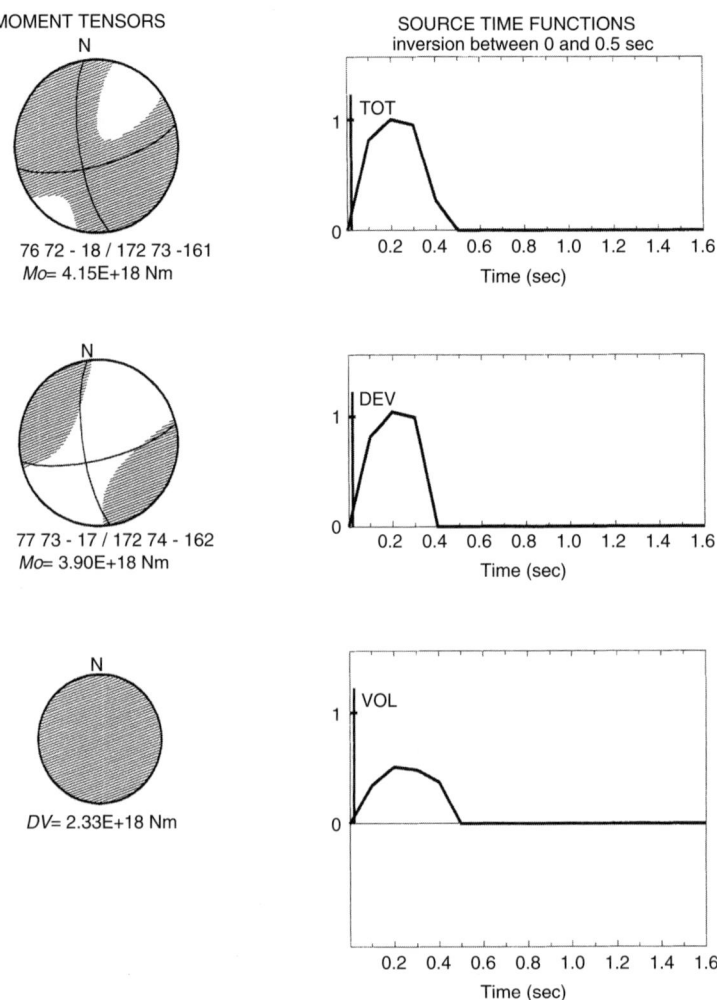

FIG. 13. Factorization of the first subevent of the 1996 large seismic event in the closed potash mine of Teutschenthal, Germany: mechanism and source time function for the total (top), the deviatoric part (middle), and the positive isotropic significant part (bottom) of the source.
Source: Reprinted from Teyssoneyre et al. (2002, Fig. 7) with kind permission from Birkhäuser Verlag AG.

site and recording instrument response. A deconvolution technique based on Empirical Green's Function (EGF) became popular in earthquake studies a long time ago. For two seismic events of different strengths, located close to each other and having similar focal mechanisms, the record of the smaller event can be considered as the EGF (Hartzell, 1978). It can be deconvolved from the record of the larger event to obtain a Relative

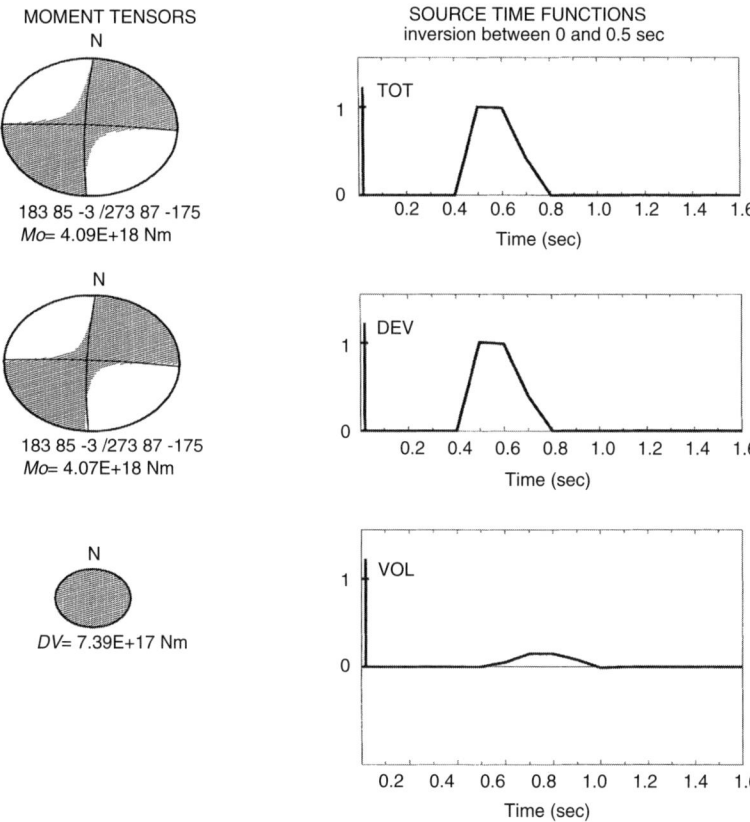

FIG. 14. Factorization of the second subevent of the 1996 large seismic event in the closed potash mine of Teutschenthal, Germany: mechanism and source time function for the total (top), the deviatoric part (middle), and the isotropic part (bottom) of the source.
Source: Reprinted from Teyssoneyre et al. (2002, Fig. 8) with kind permission from Birkhäuser Verlag AG.

Source Time Function (RSTF) at a given station for the larger event (Mueller, 1985). Such deconvolution should result in a RSTF that is corrected for the path, site and instrumental effects.

This approach is especially convenient for studying seismic events in mines, where underground seismic networks situated in the source area are often composed of a large number of sensors, and where seismic events are located with high accuracy. Surprisingly, only a few papers describing the EGF based STF of seismic events induced by mining have been published, in contrast to natural earthquakes for which tens of publications are available. Domański and Gibowicz (2001, 2003) and Domański et al. (2001, 2002a,b) described the first results from the application of the EGF technique to recover the STF of larger seismic events recorded at Rudna mine in Lubin Mining District, Poland. Seismicity there is monitored by an underground seismic network composed of 32 vertical sensors located at depths from 550 to 1150 m and extending over an area of approximately 10 by 10 km.

Altogether 43 larger events with moment magnitude ranging from 2.1 to 3.6 were studied for which the records of some 50 smaller events from the same area and with similar source mechanism were accepted as EGSs. In most cases, a double number of the records of selected event was available from two similar underground seismic networks that are in operation at Rudna and at adjacent Polkowice-Sieroszowice mines. The records of P waves from 10 up to 42 seismic stations were selected. Two methods for the computation of STF were used simultaneously for each seismic event. In the classic frequency-domain approach developed by Mueller (1985), the EGF is deconvolved by spectral division from the record of the larger event to obtain a RSTF at a given station for the larger event. The other effective time-domain approach is the projected Landweber deconvolution method which is an iterative nonlinear technique allowing the introduction of physical constraints on the final STF (Bertero et al., 1997, 1998; Piana and Bertero, 1997). This technique successfully overcomes the instability effects of the deconvolution process inherent in the frequency domain and provides stable and reliable RSTFs retrieved at several stations. The results obtained in the frequency and time domains were found to be to some extent similar, but the time-domain approach provides more objective determination of STF duration, essential for a proper determination of the source dimension. A comparison of the values of source duration derived by the spectral division (SPD) and by the Landweber deconvolution (PLD) for 43 selected events from Rudna mine is shown in Fig. 15, reprinted from Domański and Gibowicz (2003). The source duration estimated in the frequency domain is slightly higher than that estimated in the time domain.

After retrieving RSTFs from the records of several stations, it is possible to consider their dependence on the station azimuth with regard to the source location. If the pulse widths or the pulse maximum amplitudes depend on the station azimuth in a regular manner, then it is an indication that the rupture propagated unilaterally, and the rupture velocity and the rupture propagation direction can be determined simultaneously by a least-squares method either from the distribution of the RSTF pulse widths (e.g. Li and Thurber, 1988) or from the distribution of their maximum amplitudes (Li et al., 1995). Out of 43 selected events, 32 events (75 percent) displayed directivity effects, implying unilaterally propagating ruptures (Domański and Gibowicz, 2003). The rupture velocity ranged from 0.25 to 0.9 of the shear wave velocity. Its values can be divided into two distinct sets: low velocity values in comparison with those from natural earthquakes, between 0.25 and 0.6 of the shear wave velocity, and high velocity values greater than 0.6 of the shear wave velocity. The ratio of rupture velocity to the shear wave velocity displayed from its lowest to its highest values is shown in Fig. 16, reprinted from Domański and Gibowicz (2003).

Domanski and Gibowicz (2008) estimated source parameters in the time domain of 32 previously studied and 5 new seismic events, recorded at Rudna copper mine, from their RSTFs that displayed unilateral rupture propagation. They found that rupture velocity in the source is almost independent of seismic moment. The fault length, on the other hand, estimated from the average source pulse width and rupture velocity, is distinctly dependent on seismic moment and is smaller than the source radius estimated from the corner frequency on the average by about 25 percent. The static stress drop estimated in

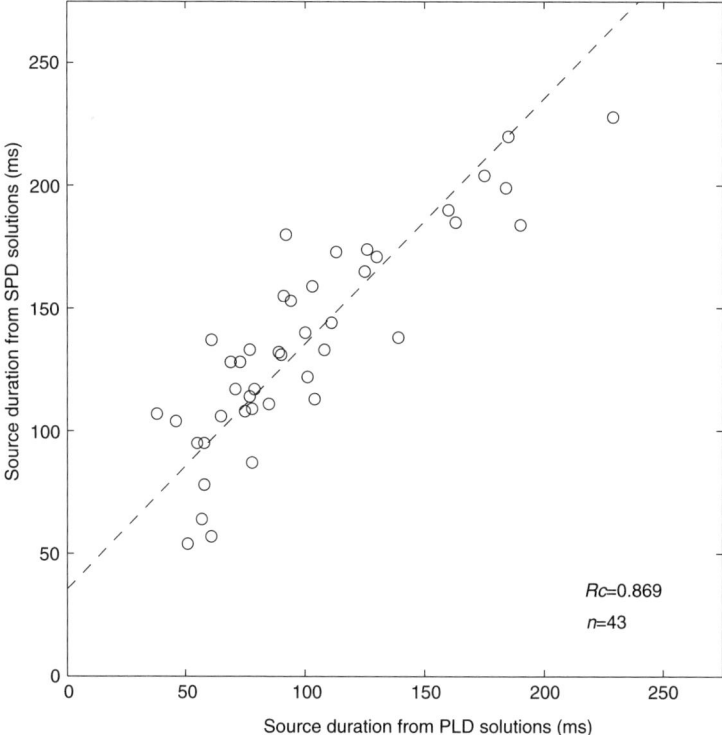

FIG. 15. Comparison of the values of source duration (pulse width) derived by the spectral division (SPD) and by the Landweber deconvolution (PLD) for seismic events from the Rudna copper mine, Poland. Their approximation is shown by a dashed line; the correlation coefficient Rc and the number of observations n are also given.
Source: Reprinted from Domański and Gibowicz (2003, Fig. 3).

the time domain is in general of the same size as that of the stress drop estimated in the frequency domain. The apparent stress increases with increasing rupture velocity, and the ratio of apparent stress to static stress drop, the radiation efficiency, seems to depend on the rupture velocity as well.

Other time-domain techniques for the determination of STF of seismic events were also proposed. Dębski and Domański (2002) discussed the application of the pseudo-spectral method to retrieve STFs in a framework of the EGF approach. The method is based on a decomposition of the searched STF into properly chosen base functions with decomposition coefficients estimated by an optimizer founded on a genetic algorithm. The method, being essentially nonlinear, was compared with the projected Landweber technique by applying both these methods to a pair of seismic events selected from Rudna mine. The first results show that the pseudo-spectral method not only leads to smoother solutions due to the implicit restriction of spectra of the STFs but also provides a better fit to the observed waveforms than the projected Landweber technique. An example of the STF calculated by the Landweber deconvolution and by the pseudo-spectral method

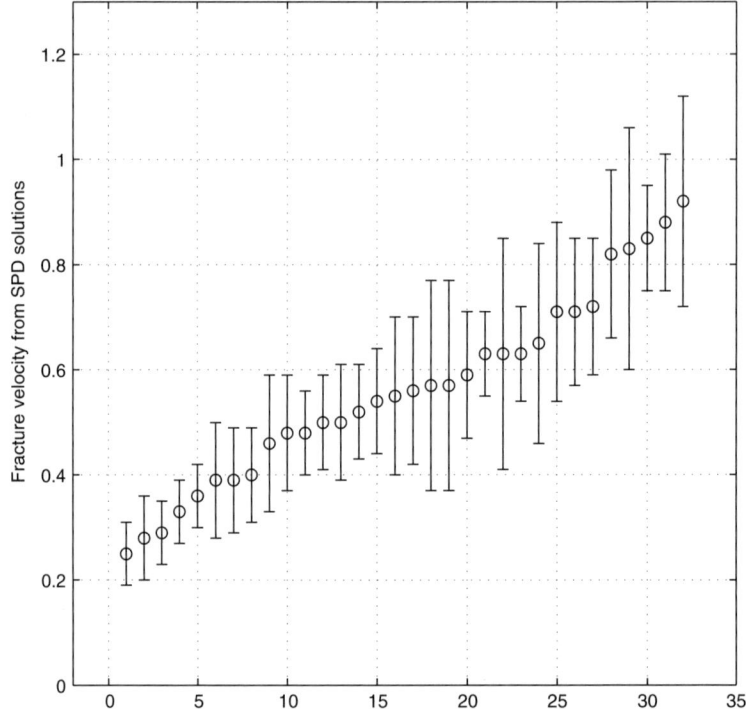

FIG. 16. The ratio of rupture velocity to the shear wave velocity displayed from its lowest to its highest values derived by the spectral division (SPD) for seismic events from the Rudna copper mine, Poland. The standard errors of rupture velocity are marked by vertical bars.
Source: Reprinted from Domański and Gibowicz (2003, Fig. 9).

for a seismic event recorded at Rudna mine is shown in Fig. 17, reprinted from Dębski and Domański (2002).

Makowski (2003) proposed a new method for the blind deconvolution of seismic signals recorded at Rudna mine. The method consists of two major steps: estimation of the rock mass impulse response (Green's function) through approximation of the signal, and inverse filtering combined with optimization. The model of seismic signals is based on the following assumptions: the source signal is a short-term signal, the rock mass constitutes a parallel connection of elementary systems, and the elementary systems are of resonant type. The application of the proposed deconvolution method in practice confirmed its adequacy for the local conditions of the rock mass and for the generated seismic events.

6. Source Parameters and Their Scaling Relations

There are two fundamental measures of earthquake size: the seismic moment, a static measure, and radiated seismic energy, a dynamic measure (e.g. Venkataraman *et al.*, 2006). These two parameters provide the apparent stress, one of the most common

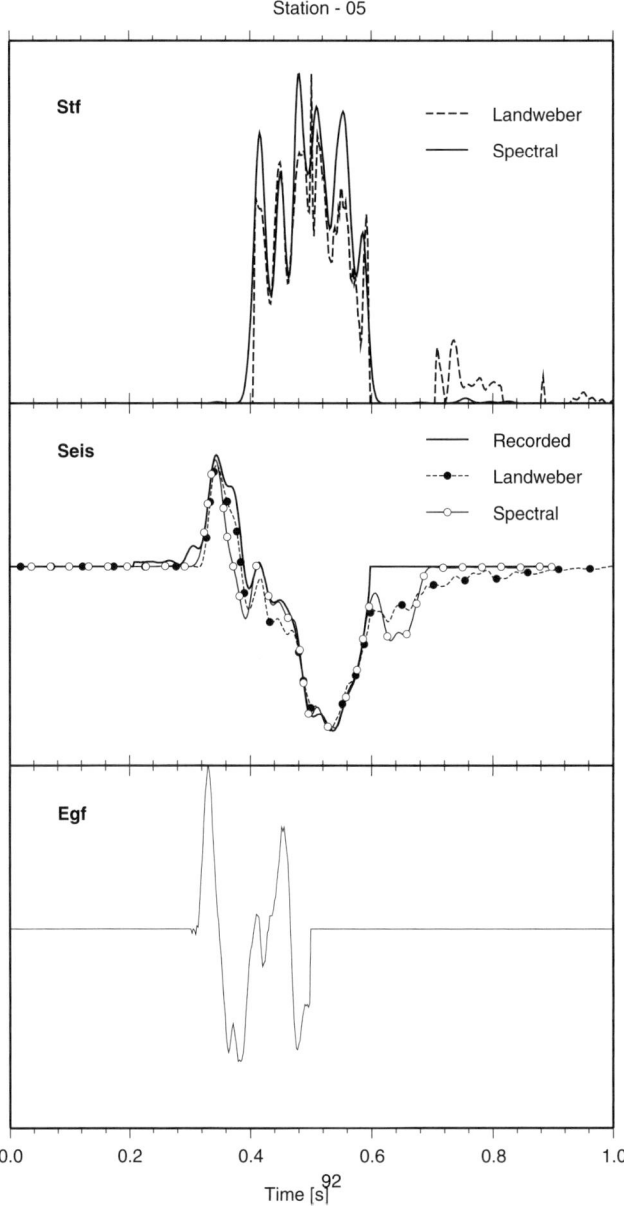

FIG. 17. The Source Time Functions (STFs) calculated by the Landweber deconvolution and by the pseudo-spectral method for a seismic event recorded by station 5 at the Rudna copper mine, Poland. The middle panel shows the observed and synthetic waveforms (Seis) obtained by the convolution of the calculated source time functions with the empirical Green's function (EGF) shown in the bottom panel.
Source: Reprinted from Dębski and Domański (2002, Fig. 4).

measures of stress release during earthquakes. The apparent stress, the ratio of radiated energy to seismic moment multiplied by the rigidity modulus (Wyss and Brune, 1968), corresponds to radiated energy per unit area and per unit slip on the fault plane. The computation of seismic energy usually requires an integration of radiated energy flux from ground velocity-squared digital seismograms, corrected for attenuation and site effects. Different techniques used to estimate the radiation energy were recently reviewed by Venkataraman *et al.* (2006). The computation of seismic moment requires in turn the spectral analysis of selected waveforms from a given seismic event. Both these measures of earthquake size are used in studies of seismicity induced by mining.

The static stress drop, the other common measure of stress release, is defined as the difference between the average shear stress in the source area before and after the earthquake. It can be expressed in terms of seismic moment and a characteristic dimension of the fault zone. The seismic moment can be reliably determined from the low frequency level of the far-field displacement spectra. For large earthquakes, the spatial distribution of their aftershocks provides an estimate of the size and shape of the rupture area, leading to a fairly reliable estimate of the static stress drop averaged over the entire fault plane. This is not the case for small seismic events, such as those generated in mines. The most commonly used estimate of static stress drop follows the source model – either of Brune (1970) or of Madariaga (1976) – and is based on the seismic moment and the fault radius represented by a circular dislocation.

The static stress drop is generally constant for earthquakes of different sizes; this self-similar scaling goes down to a moment magnitude close to zero (e.g. Abercrombie, 1995; Imanishi *et al.*, 2004; Stork and Ito, 2004; Allmann and Shearer, 2007), while the apparent stress is thought by some to increase with seismic moment (Abercrombie, 1995; Izutani and Kanamori, 2001). Further studies based on seismic records from boreholes and mines had also suggested an increase in average values of apparent stress with seismic moment (Prejean and Ellsworth, 2001; Richardson and Jordan, 2002; Venkataraman *et al.*, 2006). Beeler *et al.* (2003), on the other hand, found that variations of apparent stress with seismic moment for the Cajon Pass, California, earthquakes, recorded at 2.5 km depth, are associated with systematic variations in static stress drop with seismic moment; apparent stress and stress drop co-vary. The data used by Abercrombie and Rice (2005) over a wide magnitude range, from 0 to 7, suggest also a scale dependence of apparent stress and stress drop; both may increase slightly with earthquake size.

In contrast, McGarr (1999) found that the maximum values of apparent stress are constant over 14 orders of seismic moment. He suggested that the average values of selected data sets are affected by limited frequency-band recording. Ide and Beroza (2001) studied the effects of limited bandwidth recording on various data sets and found that they led to substantial underestimates of the apparent stress. They concluded that the corrected values of apparent stress are almost constant over 17 orders of seismic moment. McGarr and Fletcher (2003) demonstrated that over a broad range of seismic moments, apparent stress does not show any systematic dependence on seismic moment, although the individual data sets show obvious scaling, mostly related to band-limited recording. Ide *et al.* (2003) used the spectral ratio technique to re-analyze small earthquakes recorded at 2 km depth in Long Valley Caldera, California, studied previously by Prejean and Ellsworth (2001). Their measurements follow constant stress drop and constant apparent stress scaling. Static stress drop and apparent stress of microearthquakes at

Parkfield, California, are also independent of seismic moment in agreement with scaling laws reported for moderate and large earthquakes (Imanishi and Ellsworth, 2006).

Oye et al. (2005) compared source parameters of seismic events induced by mining within the Pyhäsalmi mine, Finland, derived from constant attenuation models with source parameters obtained using the multiple empirical Green's function technique. They found that the apparent stress derived using constant attenuation models seems to increase with magnitude, whereas the results based on the second approach imply constant apparent stress within the considered magnitude range. Yamada et al. (2005a,b) studied slip distributions of five small seismic events in a South African gold mine and found that their apparent stress does not decrease with decreasing seismic moment and is similar to that of larger natural earthquakes. In their recent study of twenty small seismic events in the same Mponeng gold mine, South Africa, Yamada et al. (2007) again found that both the apparent stress and static stress drop are independent of seismic moment and that dynamic rupture processes of these small events are similar to those of larger earthquakes.

The relation between seismic moment and corner frequency is the most often used scaling dependence in earthquake studies, and the seismic moment is often found to be inversely proportional to the cubed corner frequency. This does not necessarily mean that the static stress drop is constant and independent of earthquake size, although such scale independence is often implied (Kanamori and Rivera, 2004). For the proper determination of static stress drop, the rupture velocity in the source area must be known, which is one of the most difficult source parameters to estimate. Similar scaling relations are observed for seismic events in mines, even for the smallest events (e.g. Ogasawara et al., 2002c). The relation between the radiated energy and the seismic moment is the other important scaling relation, but these two relations are not entirely independent (Kanamori and Rivera, 2004).

It is often assumed that the energy of S waves comprises nearly all of the energy radiated by earthquakes, because the energy of P waves is small in comparison with that of S waves (e.g. Izutani and Kanamori, 2001; Kanamori and Rivera, 2004). This is not necessarily the case for small events induced by mining (e.g. Gibowicz et al., 1991; Urbancic and Young, 1993; Urbancic and Trifu, 1996), and their radiated energy is usually estimated as the sum of P-wave and S-wave energies. The values of radiated energy and seismic moment are usually bounded by the lines of constant apparent stress ranging from about 0.01 to a few MPa (e.g. Gibowicz, 2001; Richardson and Jordan, 2001). A least-squares linear regression performed by Gibowicz (2001) on separate data sets from seven mines showed that the slope coefficient in all cases was consistently close to 1.5, the value previously also found in some South African gold mines (e.g. Mendecki, 1993). Senatorski (2001) considered the energy-moment relations for global and regional data and proposed a statistical relationship between the radiated energy, seismic moment and source dimension in which the radiated energy is proportional to the seismic moment to power 1.5, and which is consistent with the global and regional observations. The relation can be explained by a fault zone model based on the slip-dependent constitutive law and the overdamped dynamics approximation.

The apparent stress as a function of seismic moment for seismic events from various mines in Canada, South Africa, Finland, and Poland is shown in Fig. 18. The figure is based on the eight available data sets from the Underground Research Laboratory,

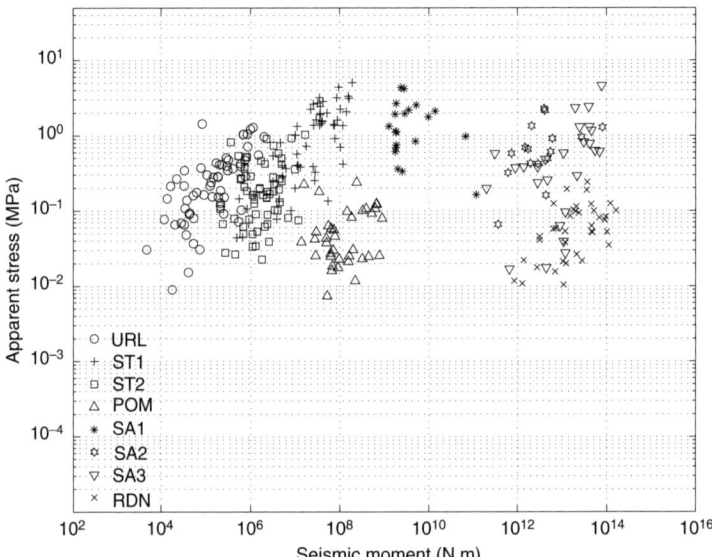

FIG. 18. Apparent stress versus seismic moment for seismic events from the Underground Research Laboratory URL, Canada (Gibowicz et al., 1991), Strathcona mine ST1, Canada (Urbancic and Young, 1993), Strathcona mine ST2 (Urbancic and Trifu, 1996), Pyhäsalmi ore mine POM, Finland (Oye et al., 2005), Mponeng gold mine SA1, South Africa (Yamada et al., 2007), South African gold mines SA2 (McGarr, 1994), South African gold mines SA3 (Richardson et al., 2005), and Rudna copper mine RDN, Poland (Domanski and Gibowicz, 2008).

Canada (Gibowicz et al., 1991), Strathcona mine, Canada (Urbancic and Young, 1993; Urbancic and Trifu, 1996), Pyhäsalmi ore mine, Finland (Oye et al., 2005), Mponeng gold mine, South Africa (Yamada et al., 2007), South African gold mines (McGarr, 1994; Richardson et al., 2005), and Rudna copper mine, Poland (Domanski and Gibowicz, 2008). Limited bandwidth recording of seismic waves can substantially influence the estimation of source parameters, such as corner frequency, static stress drop and radiated energy. Di Bona and Rovelli (1988) and Ide and Beroza (2001) demonstrated that radiated energy can be especially severely underestimated when adequate high frequencies are not recorded. They introduced corrections to account for the finite bandwidth effects and these corrections were applied here to the estimates of corner frequency, static stress drop and radiated energy, whenever needed. The apparent stress shows no systematic dependence on seismic moment over its 10 orders of magnitude from about 10^4 to 10^{14} N m, though the individual sets of observations exhibit an obvious increase of apparent stress with the size of seismic events. This could be due to variations in slip distribution over the fault plane (McGarr, 1994). Scaling relations for the apparent stress as a function of seismic moment, average slip acceleration and rupture area were proposed by Senatorski (2007) to explain statistical trends that characterize different data sets displayed on the apparent stress and seismic moment diagrams. The scaling relations are interpreted in terms of the apparent stress minimum, and although the apparent stress

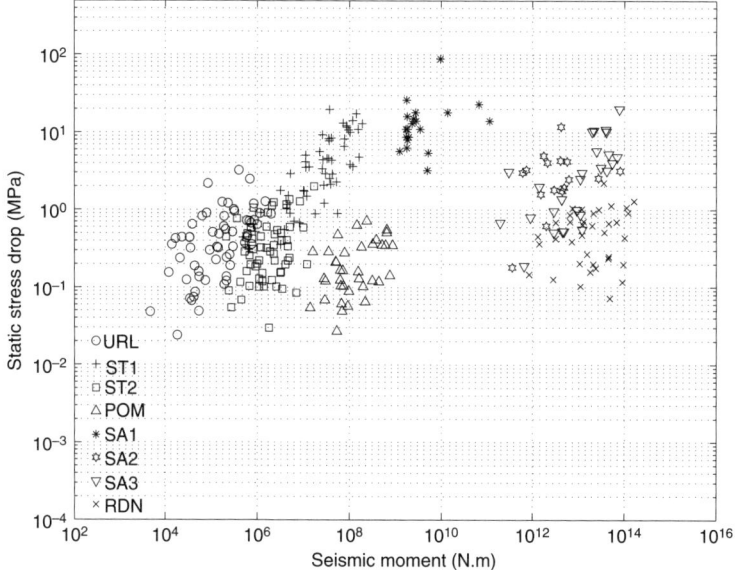

FIG. 19. Static stress drop versus seismic moment for seismic events from the Underground Research Laboratory URL, Canada (Gibowicz et al., 1991), Strathcona mine ST1, Canada (Urbancic and Young, 1993), Strathcona mine ST2 (Urbancic and Trifu, 1996), Pyhäsalmi ore mine POM, Finland (Oye et al., 2005), Mponeng gold mine SA1, South Africa (Yamada et al., 2007), South African gold mines SA2 (McGarr, 1994), South African gold mines SA3 (Richardson et al., 2005), and Rudna copper mine RDN, Poland (Domanski and Gibowicz, 2008).

on the average increases with seismic moment, the rupture processes of small and large earthquakes were found to be similar.

The static stress drop as a function of seismic moment for the same seismic events is shown in Fig. 19. The static stress drop for the events from Rudna copper mine was estimated by Domanski and Gibowicz (2008) from the relation expressing the fault length in terms of seismic moment and static stress drop (Beresnev, 2001), and the fault length was estimated in the time domain from the source time function. Similarly as in the case of apparent stress, the static stress drop displays no systematic dependence on seismic moment, but the individual sets of observations from various mines show an increase of static stress drop with the event size. From Fig. 18 and Fig. 19, it follows that both the measurements of stress release co-vary. Interestingly, the static stress drop of tiny events from acoustic emission experiments is on the average of a similar value as that found from seismic events in mines. The static stress drop as a function of seismic moment for acoustic emission events and seismic events from a deep gold mine in South Africa is shown in Fig. 20, and the seismic moment against the source radius for the same acoustic emission and seismic events is presented in Fig. 21, reprinted from Sellers et al. (2003), where the straight line, connecting both the data sets, corresponds to the scaling of seismic moment being proportional to the cubed source radius.

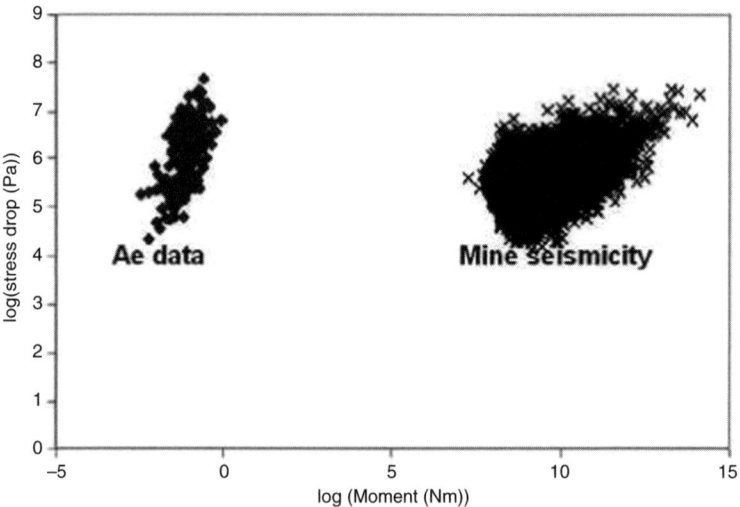

FIG. 20. Static stress drop versus seismic moment for acoustic emission events (diamonds) and seismic events (crosses) from 5 years of mining at a site on a deep-level gold mine in South Africa.
© 2003, American Geophysical Union
Source: Reprinted from Sellers et al. (2003, Fig. 10).

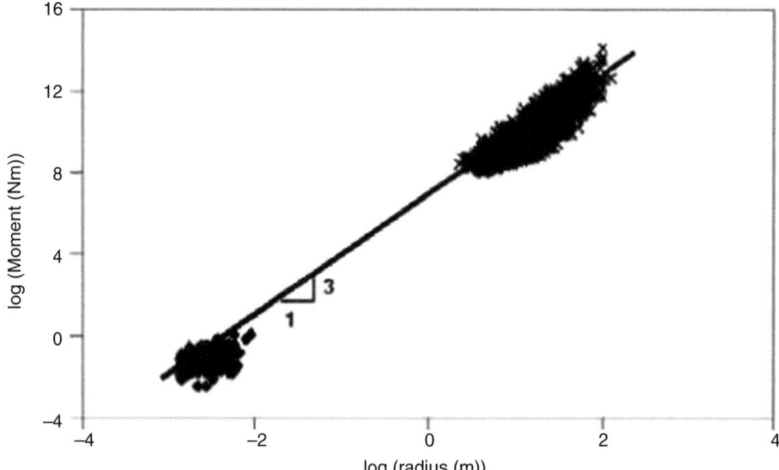

FIG. 21. Seismic moment versus source radius for acoustic emission events (diamonds) and seismic events (crosses) from 5 years of mining at a site on a deep-level gold mine in South Africa. The straight line corresponds to the scaling of seismic moment being proportional to the cubed source radius.
© 2003, American Geophysical Union
Source: Reprinted from Sellers et al. (2003, Fig. 11).

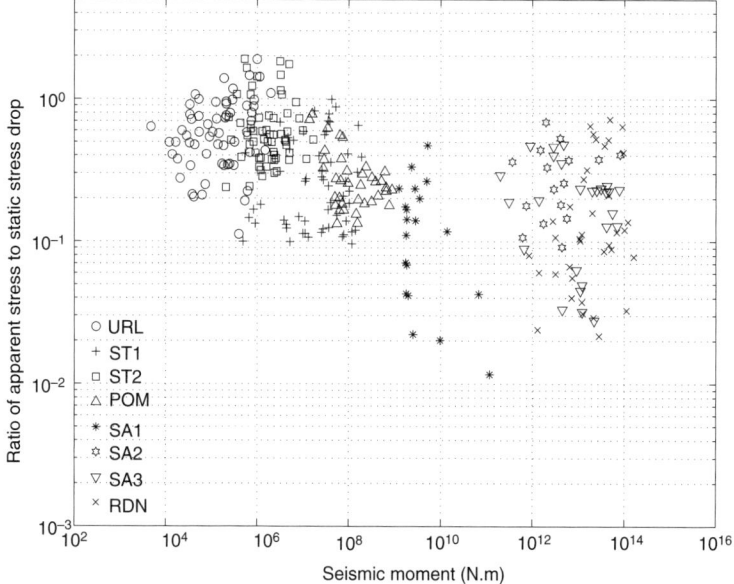

FIG. 22. Ratio of apparent stress to static stress drop versus seismic moment for seismic events from the Underground Research Laboratory URL, Canada (Gibowicz et al., 1991), Strathcona mine ST1, Canada (Urbancic and Young, 1993), Strathcona mine ST2 (Urbancic and Trifu, 1996), Pyhäsalmi ore mine POM, Finland (Oye et al., 2005), Mponeng gold mine SA1, South Africa (Yamada et al., 2007), South African gold mines SA2 (McGarr, 1994), South African gold mines SA3 (Richardson et al., 2005), and Rudna copper mine RDN, Poland (Domanski and Gibowicz, 2008).

Although the conventional seismic efficiency, the ratio of radiated energy to total released energy, or equivalently the ratio of apparent stress to mean stress, has been estimated for some earthquakes (e.g. Savage and Wood, 1971; McGarr, 1994, 1999), the mean stress in the source area cannot be determined from seismic data and is usually not well known. A different measure of efficiency, the ratio of apparent stress to static stress drop can be determined from seismic observations, and is called by Beeler et al. (2003) the Savage-Wood efficiency, because Savage and Wood (1971) first considered how apparent stress is related to static stress drop. A quantity equal to twice the Savage-Wood efficiency is called either the radiation efficiency (Husseini and Randall, 1976) or the apparent radiation efficiency (Venkataraman et al., 2006). The ratio of apparent stress to static stress drop as a function of seismic moment for the eight data sets from seismic events recorded in various mines is shown in Fig. 22. In general, the ratio seems to be independent of seismic moment, except that it displays low values at the seismic moment close to 10^{10} N m (moment magnitude of 0.6), where they are distinctly lower than the others, though their scatter is considerable. This specific value of seismic moment corresponds to its "border" value that divides type A events from type B events of Richardson and Jordan (2001, 2002). The radiation efficiency could possibly be also interpreted as showing a slight tendency of increasing with decreasing seismic moment, though the scatter of the observed values is considerable, possibly indicating the range of errors involved in the determination of the stress release parameters. Savage and Wood (1971) demonstrated that the apparent stress has an upper bound of 1/2 the static stress drop and several of our observations exceed this limit.

7. PRECURSORY PHENOMENA AND PREDICTION OF LARGE SEISMIC EVENTS

The Japanese research group has initiated investigation of earthquake-generation processes in South African deep gold mines, where mining takes place at depths of 2000-3000 m and seismic events are induced in the close vicinity of stopes. Seismogenic processes therefore can be monitored at very short distances with sensors installed ahead of time in seismogenic areas, and this approach is called a semi-controlled earthquake-generation experiment (Ogasawara *et al.*, 2001, 2002b, 2005a,b). With the use of modern technology developed in Japan and South Africa, Japanese and South African seismologists have been collaborating since 1992 to understand the entire life span, all source processes of an earthquake. They installed a sensitive and stable strainmeter in a potential seismic source area at Bambanani mine, Welkom, and successfully monitored strain changes with a resolution of 24 bit and sampling frequency of 25 Hz during the life span of two seismic events with magnitude greater than 2, at a depth of some 2.5 km and at distances less than 100 m from the instrument (Ogasawara *et al.*, 2005a). The associated strain changes were up to $\sim 10^{-4}$ which corresponds to a stress change of about 7 MPa. Concentrated foreshocks with significant strain steps and logarithmic post-seismic deformation were recorded during several tens of seconds before one seismic event. Monotonic relaxation of maximum principal strain was observed during several days before both the events, but it was unclear whether this relaxation took place in the source area. An array of multiple strainmeters was needed and was deployed at Tau Tona and Mponeng gold mines during the next stage of the study between 2003 and 2004 (Ogasawara *et al.*, 2005b; Naoi *et al.*, 2006).

During semi-controlled earthquake-generation experiments conducted in 1996 at a depth of 2.6 km from the surface at a gold mine in South Africa, described in some detail by Ogasawara *et al.* (2002b), a search for initial phases, also called seismic nucleation phases or slow initial phases, was undertaken and the results were described by Satoh *et al.* (2002). Altogether some 20,000 seismic events were recorded at distances shorter than 200 m (the nearest event was located about 10 m from a seismometer) by a high sampling rate and a wide dynamic range seismic system. The small initial phases, preceding large main *P*-wave first arrivals, were sought on the seismograms of the largest events and were found on the records of all six such events with seismic moments that ranged from 5.0×10^{10} to 1.4×10^{11} N m. The duration of these small initial phases was comparable to that reported for natural earthquakes of similar strength and recorded at distances of several kilometers. The waveforms of initial phases observed in deep mines, on the other hand, are rather complex, resembling those of the moderate and large natural earthquakes. A search for the initial phases generated by the seismic events at Rudna copper mine in Poland was carried out by Kwiatek (2003). Several thousand records of 137 seismic events with seismic moments between 2.7×10^{11} and 3.4×10^{14} N m were analyzed. The initial phase was found on 182 records, visible as a gradual and slow increase of the signal amplitude directly before the *P*-wave first arrival. The waveforms were corrected for the effects of the recording system, the attenuation and the noise. It seems that the initial phases originated in the seismic source and were not an effect connected with the wave propagation from the source to the station. Its duration depends on the seismic moment.

A stress monitoring site for stress-forecasting earthquakes, based on shear wave splitting, was described by Crampin and Jackson (2001) with its possible applications to seismicity induced by mining. Almost all rocks display seismic shear wave splitting with stress-aligned polarizations. Theory and observations show that splitting is controlled by small-scale deformation of the distributions of fluid saturated stress-aligned grain-boundary cracks and low aspect-ratio pores pervading almost all rocks. Since this microcrack geometry is the most compliant component of the rock mass, shear wave splitting directly monitors the response of rock mass to small changes of stress. In the past, it had been assumed that the accumulation of stress before earthquakes continued until stress was released by faulting at the time of the earthquake. New data and re-appraisal of existing data now, however, suggest that the stress begins to relax, and cracks close from tens of minutes to months before the earthquake actually occurs, with the logarithm of the relaxation duration proportional to the magnitude of the impending earthquake (Gao and Crampin, 2004). The duration of the relaxation appears to be directly correlated with earthquake magnitude, and may have implications for the earthquake source processes and the ability to predict earthquakes.

An attempt to monitor stresses in the rock mass by a shear wave splitting analysis in underground mines was undertaken in 1996 during the semi-controlled earthquake-generation experiments at Western Deep Levels gold mine in South Africa, described by Nagai et al. (2002). Two thin veins of gold reefs were mined there at depths of 2.6 and 3.6 km. The observation site was located in a tunnel at a depth of 2650 m, 50 m beneath one active stope and 1 km above another active stope. Shear wave splitting was clearly observed, and it was found that the anisotropic area was located near the seismic sources around the deeper stope at 3.6 km. The area near the seismic stations was isotropic or weakly anisotropic. The time difference between two split shear waves was larger for deeper seismic events and it decreased with the progress of mining activity. The crack density became higher during the stress concentration under the influence of mining and it became lower after the occurrence of seismic events resulting in stress release.

The use of borehole deformation measurements for the evaluation of rock mass conditions in copper mines in Poland was described by Butra and Orzepowski (2001). Between 1996 and 2000, specific changes of deformation in vertical boreholes were investigated at Rudna mine as precursors to larger seismic events that occur in the mine. It was found that the changes of deformation before the large seismic events were observed simultaneously in most boreholes located over a large area in the roof strata. During the period of monitoring, 15 large seismic events out of 16 were preceded by the precursors in at least two boreholes simultaneously, 6 days on the average before the event. Several precursory changes, however, not followed by large seismic events were also observed. The results of similar measurements of borehole deformations in coal mines in the Upper Silesian Coal Basin, Poland, were described by Matwiejszyn and Ptak (2002) and Lasek et al. (2005).

Mansurov (2001) proposed a method for the prediction of the occurrence of strong seismic events based on the kinetic conception of the strength of solids and a rigid discontinuity model. The analysis of the space and time structure of seismicity induced by mining allows us to establish the most general rules of the evolution of rock failure processes and to define the most informative prediction characteristics, such as the concentration parameter, volume concentration, average size of faults, and others. Retrospective predictions applied to seismic events in the North Ural Bauxite Mines showed that this technique is efficient and can be used in practice. A similar approach to the prediction of large seismic events at Champion Reef Mine of the Kolar gold fields in southern India was described by Srinivasan et al. (1999). Microseismic emission generated by deformation and cracking of the rock mass around stopes under changing stresses was found to provide useful information on impending major seismic events. Typical characteristics of the space-time patterns of microseismic emission were shown to be of diagnostic value for the prediction of large events that follow such emission. Three reliable short-term precursors were found: the rate of variation of the number of small events, their radiated energy, and predominant frequency of their waveforms. A distinct and rapid increase in the microseismic event rate in comparison with the normally fluctuating background activity usually preceded a major event. The radiated energy tended to increase before the occurrence of a large seismic event. A marked spectral shift towards lower predominant frequency on microseismic waveforms indicated an imminent major event.

The space-time correlation of microseismicity recorded in 1997-1998 at Creighton Mine, Ontario, Canada, was analyzed by Marsan et al. (1999, 2000). They identified and quantified a stress diffusion mechanism, and showed that the spatial distribution of the aftershocks correlated with a main event occurring at t after the main event changes with t. This change takes the form of an expanding pattern, characterized by a typical scale varying as t^H, where H is the diffusion exponent estimated to be 0.18. This diffusion exponent was found to increase when considering only the subset of the largest events as main shocks. These observations could be explained by modeling the seismicity system as a scale-invariant set of connected faults on which individual stress concentrations undergo spatial relaxations. The increase of H with the increase in energy release of the main event can be interpreted as the indication of a stress diffusion mechanism, involving propagation on the heterogeneous fractal fault network.

Ouillon and Sornette (2000) described tests – performed on seismic events recorded in 1997 in a deep South African gold mine in the Welkom area – of the concept that large earthquakes are "critical points". They extended the concept of an optimal time and space correlation region and tested it on the eight largest seismic events out of 2487 events listed in their catalog. In the first test, they used the simple power-law time-to-failure formula. In spite of the fact that the search for the optimal correlation size was performed with this simple power law, strong evidence for accelerated seismicity and for the presence of log-periodic behavior with a scaling factor close to 2 was found. This led to a new algorithm based on a space and time smoothing procedure, which was also intended to account for the finite range and finite duration of mechanical interactions between the events. The new algorithm was preliminarily tested on the largest seismic events only. A high-quality result was obtained, with considerable improvement in accuracy and robustness. This confirms the potential importance of log-periodic signals. The study indicates the way for an efficient implementation of systematic testing procedure for real time predictions of large seismic events. The concept of "critical rupture" and "finite-time singularity", and its application to seismicity in mines were reviewed by Sornette (2001).

Reliable predictions regarding the location and magnitude of an impending rockburst were demonstrated by Wiles (2005) who used a calibrated numerical model. Although failure was imminent, the prediction of its exact time was very uncertain. Back-analysis showed that prediction can be made with an uncertainty of less than 10 percent in terms of stress. Using this figure, a prediction with 90 percent confidence was made that the crown pillar at Creighton Mine, Ontario, Canada, would fail when its width is 33 ± 18 m, an uncertainty representing well over half of the pillar width. The uncertainty in terms of stress was greatly amplified in application to estimate the pillar width at the moment of failure. Higher accuracy predictions could be obtained with lower uncertainties. But even by using improved geological details and more complex modeling capabilities, it seems unlikely that the values of uncertainty in terms of stress significantly smaller than 10 percent could ever be attained.

Using numerical modeling, Cai *et al.* (2005) investigated the time-space-strength relations for seismicity induced by mining at depths greater than 900 m at Laohutai coal mine in Fushun, Liaoning Province, north-eastern China. There are three main types of source mechanisms: shearing on faults, gravity impact of roof collapses, and gravity impact of pillar collapses. The strongest seismic events are generated by movement on faults. The numerical model of mining disturbance potential led to the quantitative relations between the seismic event and a number of quantities such as mining depth, mining volume, excavation methods, tectonic structure, and stress conditions. The model provided an effective method for mining design and safer production.

Thus, similarly to some extent as for natural earthquakes, no consistently reliable phenomena precursory to large seismic events induced by mining have been found. Nevertheless, several retrospective predictions and a few predictions in real time have been made and described. In contrast to natural seismicity, prevention is the most important approach to seismicity in mines. The development and improvement of methods for mine design to reduce large-magnitude seismic events is one of the major challenges for mining seismicity and rock mechanics studies. Modern numerical modeling of rock mass response to mining provides proper technology to solve these problems.

8. Summary

1. From the beginning of the twenty first century, a number of new techniques have been introduced and new significant results have been obtained in studies of seismicity induced by mining. International Symposia on Rockbursts and Seismicity in Mines (RaSiM), regularly held every fourth year in various countries, helped us to understand some fundamental ideas about the nature and causes of seismic events in mines. Recognizing that seismicity is the response of the rock mass to stress and strain changes induced by mining, implying that mine design can influence the level of seismicity, is of practical importance.
2. During the last decade, seismic monitoring has been expanded in several mining districts, especially in Australia, China and Russia. Various seismic monitoring systems are in operation at different mines and in different countries. From the

waveforms recorded by a modern digital system, the location, origin time, scalar seismic moment, radiated energy and other source parameters of a seismic event can be routinely estimated. The processing of waveforms is very laborious, but continuous improvement in automatic processing is in progress. Acoustic emission systems monitoring stress-induced microcracking in mines recently introduced advanced recording and data processing and analysis methods, similar to those used in the monitoring of lower frequency and higher energy events. This is a major development in studies of local rock mass response to excavation stresses.

3. Mine design is based on the assumption that the rock mass response to mining can be predicted with some degree of confidence. An integrated approach, based on the use of numerical modeling of the behavior of the rock mass around a mine and on the recording and analysis of seismicity there, is advocated for estimating the likely response of the rock to future mining, for predicting the amount of seismicity that will be likely to occur over the subsequent months in a given area. Various strategies are designed to reduce the occurrence of damaging seismic events and their effects in different mines. Detailed in situ studies of rockburst fractures, of their nature and geometry, have been carried out in deep gold mines in South Africa, providing significant new data.

4. The seismic source mechanism plays an important role in understanding the various modes of rock failure observed in underground mines. The moment tensor inversion technique is the most effective method in this respect and has been used extensively since the early 1990s in studies of seismicity in mines. Several fast and automatic inversion techniques have been introduced in recent years and applied in practice. The classic double-couple seismic source model is not always representative of mining environments.

5. The source time function of a seismic event describes the moment rate release and the rupture evolution in the source. The extraction of the source time function from seismograms requires separation of the source effects from those of the path, site and recording instrument response. A deconvolution technique based on empirical Green's function, which is the record of a smaller event located close to the main event and having a similar focal mechanism, is especially convenient for studying seismic events in mines. This approach has been applied in studies of some 40 seismic events at the Rudna copper mine, Poland. Directivity effects, implying unilaterally propagating ruptures, were observed in most cases, and rupture velocity and rupture direction were estimated. The rupture velocity ranged from 0.3 to 0.9 of the shear wave velocity.

6. The apparent stress of seismic events from various mines in Canada, South Africa, Finland and Poland seems to be independent of seismic moment over its 10 orders of magnitude from 10^4 to 10^{14} N m. The static stress drop shows a similar pattern, with the exception of its few values at seismic moment close to 10^{10} N m (magnitude of 0.6), where they are distinctly higher than the others. As a consequence, the radiation efficiency, defined as twice the apparent stress divided by the static stress drop, displays low values for the same seismic moment, though their scatter is considerable.

7. Strain changes at distances less than 100 m from the sources of two seismic events and concentrated foreshocks with strain steps and logarithmic post-seismic

deformation were observed by the Japanese research group in South African deep gold mines. They also attempted to monitor stresses in the rock mass by a shear wave analysis. Shear wave splitting was clearly observed and it was found that the anisotropic area was located near the seismic sources. Specific changes of deformation in vertical boreholes in the roof strata in Polish copper and coal mines were in several cases followed by large seismic events, but contrary cases were also observed. The concepts of stress diffusion and of "critical earthquake", and numerical modeling of rock mass response provide some examples of attempted predictions. But in general, similarly as for natural earthquakes, no consistently reliable phenomena precursory to large seismic events induced by mining have been found.

Acknowledgments

I am very grateful to Art McGarr and two anonymous reviewers for their most helpful comments and suggestions which considerably improved the original manuscript, and I am especially impressed by the most detailed review prepared by Art McGarr. Woytek Debski, my colleague from the Institute of Geophysics, Polish Academy of Sciences, was most helpful in preparing the final figures in a proper electronic format and in electronic submission of the manuscript. This work was partially supported by the Polish Ministry of Science and Higher Education under contract no. 2P04D 033 30.

References

Abercrombie, R.E. (1995). Earthquake source scaling relationships from -1 to 5 m_l using seismograms recorded at 2.5 km depth. *J. Geophys. Res.* **100**, 24015–24036.

Abercrombie, R.E., Rice, J.R. (2005). Can observations of earthquake scaling constrain slip weakening? *Geophys. J. Int.* **162**, 406–424.

Adams, D.J., van der Heever, P. (2001). An overview of seismic research co-ordinated by simrac since its inception. In: van Aswegen, G., Durrheim, R.J., Ortlepp, W.D. (Eds.), Rockbursts and Seismicity in Mines. Dynamic Rock Mass Response to Mining. South African Inst. Min. Metal, Johannesburg, pp. 205–212.

Alexander, J., Trifu, C.-I. (2005). Monitoring mine seismicity in Canada. In: Potvin, Y., Hudyma, M. (Eds.), Controlling Seismic Risk. Rockbursts and Seismicity in Mines. Australian Centre for Geomechanics, Nedlands, pp. 353–358.

Allmann, B.P., Shearer, P.M. (2007). Spatial and temporal stress drop variations in small earthquakes near Parkfield, California. *J. Geophys. Res.* **112**, B04305, doi:10.1029/2006JB004395.

Andersen, L.M., Spottiswoode, S.M. (2001). A hybrid relative moment tensor methodology. In: van Aswegen, G., Durrheim, R.J., Ortlepp, W.D. (Eds.), Rockbursts and Seismicity in Mines. Dynamic Rock Mass Response to Mining. South African Inst. Min. Metal, Johannesburg, pp. 81–89.

Arabasz, W.J., Nava, S.J., McCarter, M.K., Pankow, K.L., Pechmann, J.C., Ake, J., McGarr, A. (2005). Coal-mining seismicity and ground-shaking hazard: A case study in the Trail Mountain Area, Emery County, Utah. *Bull. Seismol. Soc. Am.* **95**, 13–38, doi:10.1785/0120040045.

Baumgardt, D.R., Leith, W. (2001). The Kirovskiy explosion of September 29, 1996: Example of a CTB event notification for a routine mining blast. *Pure Appl. Geophys.* **158**, 2041–2058.

Beeler, N.M., Wong, T.-F., Hickman, S.H. (2003). On the expected relationships among apparent stress, static stress drop, effective shear fracture energy, and efficiency. *Bull. Seismol. Soc. Am.* **93**, 1381–1389.

Beresnev, I.A. (2001). What we can and cannot learn about earthquake sources from the spectra of seismic waves. *Bull. Seismol. Soc. Am.* **91**, 397–400.

Bertero, M., Bindi, D., Boccacci, P., Cattaneo, M., Eva, C., Lanza, V. (1997). Application of the projected Landweber method to the estimation of the source time function in seismology. *Inv. Prob.* **13**, 465–486.
Bertero, M., Bindi, D., Boccacci, P., Cattaneo, M., Eva, C., Lanza, V. (1998). A novel blind-deconvolution method with an application to seismology. *Inv. Prob.* **14**, 815–833.
Bowers, D., Walter, W.R. (2002). Discriminating between large mine collapses and explosions using teleseismic *P* waves. *Pure Appl. Geophys.* **159**, 803–830.
Brune, J.N. (1970). Tectonic stress and spectra of seismic shear waves from earthquakes. *J. Geophys. Res.* **78**, 4997–5009.
Butra, J., Orzepowski, S. (2001). New method of high energy event precursory symptoms detection. In: van Aswegen, G., Durrheim, R.J., Ortlepp, W.D. (Eds.), Rockbursts and Seismicity in Mines. Dynamic Rock Mass Response to Mining. South African Inst. Min. Metal, Johannesburg, pp. 535–542.
Cai, M., Kaiser, P.K., Martin, C.D. (2001). Quantification of rock mass damage in underground excavations from microseismic event monitoring. *Int. J. Rock Mech. Min. Sci* **38**, 1135–1145.
Cai, M.F., Ji, H.G., Wang, J.A. (2005). Study of the time–space–strength relation for mining seismicity at Laohutai coal mine and its prediction. *Int. J. Rock Mech. Min. Sci* **42**, 145–151.
Chen, D., Gray, L., Hudyma, M.R. (2005). Understanding mine seismicity—A way to reduce mining hazards at Barrick's Darlot Gold Mine. In: Potvin, Y., Hudyma, M. (Eds.), Controlling Seismic Risk. Rockbursts and Seismicity in Mines. Australian Centre for Geomechanics, Nedlands, pp. 269–274.
Cichowicz, A. (2005). Automatic processing and relocation methods. In: Potvin, Y., Hudyma, M. (Eds.), Controlling Seismic Risk. Rockbursts and Seismicity in Mines. Australian Centre for Geomechanics, Nedlands.
Collins, D.S., Pettitt, W.S., Young, R.P. (2002). High-resolution mechanics of a microearthquake sequence. *Pure Appl. Geophys.* **159**, 197–219.
Collins, D.S., Young, R.P. (2000). Lithological controls on seismicity in granitic rocks. *Bull. Seismol. Soc. Am.* **90**, 709–723.
Côté, M., Mitchelson, C., Alcott, J.M. (2001). Seismic monitoring and numerical modelling decision-making at McWatters' Sigma Mine. In: van Aswegen, G., Durrheim, R.J., Ortlepp, W.D. (Eds.), Rockbursts and Seismicity in Mines. Dynamic Rock Mass Response to Mining. South African Inst. Min. Metal, Johannesburg, pp. 427–431.
Crampin, S., Jackson, P. (2001). Developing a stress-monitoring site for stress-forecasting earthquakes: Possible applications to mining-induced seismicity. In: van Aswegen, G., Durrheim, R.J., Ortlepp, W.D. (Eds.), Rockbursts and Seismicity in Mines. Dynamic Rock Mass Response to Mining. South African Inst. Min. Metal, Johannesburg, pp. 133–141.
Dahm, T., Manthei, G., Eisenblatter, J. (1999). Automated moment tensor inversion to estimate source mechanisms of hydraulically induced micro-seismicity in salt rock. *Tectonophysics* **306**, 1–17.
Dębski, W. (1996). Location of seismic events—a quest for accuracy. In: Jacobsen, B.H., Moosegard, K., Sibani, P. (Eds.), Inverse Methods, Interdisciplinary elements of Methodology, Computation and Application. In: Lecture Notes in Earth Sciences, vol. 63, Springer-Verlag, Berlin, pp. 161–166.
Dębski, W. (2004). Application of Monte Carlo techniques for solving selected seismological inverse problems. *Publs. Inst. Geophys. Pol. Acad. Sc.* **B-34** (367), 1–207.
Dębski, W., Domański, B. (2002). An application of the pseudo-spectral technique to retrieving source time function. *Acta Geophys. Pol.* **50**, 207–221.
Di Bona, M., Rovelli, A. (1988). Effects of the bandwidth limitation on stress drops estimated from integrals of the ground motion. *Bull. Seismol. Soc. Am.* **78**, 1818–1825.
Domański, B., Gibowicz, S.J. (2001). Comparison of source time functions retrieved in the frequency and time domains for seismic events in a copper mine in Poland. *Acta Geophys. Pol.* **XLIX** (2), 169–188.
Domański, B., Gibowicz, S.J. (2003). The accuracy of source parameters estimated from source time function of seismic events at Rudna copper mine in Poland. *Acta Geophys. Pol.* **51** (4), 347–367.
Domanski, B., Gibowicz, S.J. (2008). Comparison of source parameters estimated in the frequency and time domains for seismic events at the Rudna copper mine, Poland. *Acta Geophys.* **56**, 324–343, doi:10.2478/s11600-008-0014-1.
Domański, B., Gibowicz, S.J., Wiejacz, P. (2001). Source time functions of seismic events induced at a copper mine in Poland: Empirical Green's function approach in the frequency and time domains. In: van Aswegen, G., Durrheim, R.J., Ortlepp, W.D. (Eds.), Dynamic Rock Mass Response to Mining. South African Inst. of Mining and Metallurgy, Johannesburg, pp. 99–108.

Domański, B., Gibowicz, S.J., Wiejacz, P. (2002a). Source parameters of seismic events in copper mine in Poland based on empirical Green's functions. In: Ogasawara, H., Yanagidani, T., Ando, M. (Eds.), Seismogenic Process Monitoring. Balkema, Lisse, pp. 75–89.

Domański, B., Gibowicz, S.J., Wiejacz, P. (2002b). Source time function of seismic events at Rudna copper mine, Poland. *Pure Appl. Geophys.* **159**, 131–144.

Dor, O., Reches, Z., van Aswegen, G. (2001). Fault zones associated with the Matjhabeng earthquake, 1999, South Africa. In: van Aswegen, G., Durrheim, R.J., Ortlepp, W.D. (Eds.), Rockbursts and Seismicity in Mines. Dynamic Rock Mass Response to Mining. South African Inst. Min. Metal, Johannesburg, pp. 109–112.

Dunlop, R., Gaete, B.S. (2001). An estimation of the induced seismicity related to a caving method. In: van Aswegen, G., Durrheim, R.J., Ortlepp, W.D. (Eds.), Rockbursts and Seismicity in Mines. Dynamic Rock Mass Response to Mining. South African Inst. Min. Metal, Johannesburg, pp. 281–285.

Dunn, M.J. (2005). Seismicity in a scattered mining environment—A rock engineering interpretation. In: Potvin, Y., Hudyma, M. (Eds.), Controlling Seismic Risk. Rockbursts and Seismicity in Mines. Australian Centre for Geomechanics, Nedlands, pp. 337–346.

Duplancic, P., Brady, B.H.G. (2001). Understanding cave behaviour through back analysis of stress, structure and microseismicity. In: van Aswegen, G., Durrheim, R.J., Ortlepp, W.D. (Eds.), Rockbursts and Seismicity in Mines. Dynamic Rock Mass Response to Mining. South African Inst. Min. Metal, Johannesburg, pp. 313–318.

Durrheim, R. (2001). Management of mining-induced seismicity in Ultra-Deep South African Gold Mines. In: van Aswegen, G., Durrheim, R.J., Ortlepp, W.D. (Eds.), Rockbursts and Seismicity in Mines. Dynamic Rock Mass Response to Mining. South African Inst. Min. Metal, Johannesburg, pp. 213–219.

Durrheim, R., Anderson, R., Cichowicz, A., Ebrahim-Trollope, R., Hubert, G., Kijko, A., McGarr, A., Ortlepp, W., van der Merwe, N. (2006). The risks to miners, mines and the public posed by large seismic events in the gold mining districts of South Africa. In: Potvin, Y., Hadjigeorgiou, J., Stacey, T. (Eds.), Challenges in Deep and High Stress Mining. Australian Centre for Geomechanics, Nedlands, Chapter 4.

Finnie, G.J. (1999). Using neural networks to discriminate between genuine and spurious seismic events in mines. *Pure Appl. Geophys.* **154**, 41–56.

Fletcher, J.B., McGarr, A. (2005). Moment tensor inversion of ground motion from mining-induced earthquakes. *Bull. Seismol. Soc. Am.* **95**, 48–57, doi:10.1785/0120040047.

Gao, Y., Crampin, S. (2004). Observations of stress relaxation before earthquakes. *Geophys. J. Int.* **157**, 578–582.

Gay, N.C., Ortlepp, W.D. (1979). Anatomy of mining-induced fault zone. *Bull. Geol. Soc. Am.* **90**, 47–58.

Gibowicz, S.J. (1990). Seismicity induced by mining. *Adv. Geophys.* **32**, 1–74.

Gibowicz, S.J. (2001). Radiated energy scaling for seismic events induced by mining. *Acta Geophys. Pol.* **49**, 95–111.

Gibowicz, S.J. (2006). Seismic doublets and multiplets at Polish coal and copper mines. *Acta Geophys.* **54**, 142–157.

Gibowicz, S.J., Lasocki, S. (2001). Seismicity induced by mining: Ten years later. *Adv. Geophys.* **44**, 39–181.

Gibowicz, S.J., Young, R.P., Talebi, S., Rawlence, D.J. (1991). Source parameters of seismic events at the Underground Research Laboratory in Manitoba, Canada: Scaling relations for events with moment magnitude smaller than -2. *Bull. Seismol. Soc. Am.* **81**, 1157–1182.

Goforth, T.T., Hetzer, C.H., Stump, B.W. (2006). Characteristics of regional seismograms produced by delay-fired explosions at the Minntac iron mine, Minnesota. *Bull. Seismol. Soc. Am.* **96**, 272–287, doi:10.1785/0120050068.

Guha, S.K. (2000). Induced Earthquakes. Kluwer Acad. Pub., Dordrecht.

Hartzell, S.H. (1978). Earthquake aftershocks as Green's functions. *Geophys. Res. Lett.* **5**, 1–5.

Hazzard, J.F., Young, R.P. (2002). Moment tensors and micromechanical models. *Tectonophysics* **356**, 181–197.

Hofmann, G., Sewjee, R., van Aswegen, G. (2001). First steps in the integration of numerical modelling and seismic monitoring. In: van Aswegen, G., Durrheim, R.J., Ortlepp, W.D. (Eds.), Rockbursts and Seismicity in Mines. Dynamic Rock Mass Response to Mining. South African Inst. Min. Metal, Johannesburg, pp. 397–404.

Holub, K. (1999). Changes in the frequency-energy distribution of seismic events during mining in the Ostrava-Karvina coal field. *Stud. Geophys. Geodet.* **43**, 147–162.

Hussein, M.I., Randall, M.J. (1976). Rupture velocity and radiation efficiency. *Bull. Seismol. Soc. Am.* **66**, 1173–1187.

Iannacchione, A.T., Burke, L.M., Chapman, M.C. (2005). Characterizing roof fall signatures from underground mine. In: Potvin, Y., Hudyma, M. (Eds.), Controlling Seismic Risk. Rockbursts and Seismicity in Mines. Australian Centre for Geomechanics, Nedlands, pp. 619–629.

Ide, S., Beroza, G.C. (2001). Does apparent stress vary with earthquake size? *Geophys. Res. Lett.* **28**, 3349–3352.

Ide, S., Beroza, G.C., Prejeanand, S.G., Ellsworth, W.L. (2003). Apparent break in earthquake scaling due to path and site effects on deep borehole recordings. *J. Geophys. Res.* **108**, 2271, doi:10.1029/2001JB001617.

Idziak, A.F. (1999). A study of spatial distribution of induced seismicity in the Upper Silesian Coal Basin. *Nat. Hazards* **19**, 97–105.

Imanishi, K., Ellsworth, W.L. (2006). Source scaling relationships of microearthquakes at Parkfield, CA, determined using the SAFOD Pilot Hole seismic array. In: Abercrombie, R., McGarr, A., Kanamori, H., Toro, G.D. (Eds.), Earthquakes: Radiation Energy and the Physics of Faulting. In: Geophys. Monogr. Ser., vol. 170, AGU, Washington DC, pp. 81–90.

Imanishi, K., Takeo, M., Ellsworth, W.L., Ito, H., Matsuzawa, T., Kuwahara, Y., Iio, Y., Horiuchi, S., Ohmi, S. (2004). Source parameters and rupture velocities of microearthquakes in western Nagano, Japan, determined using stopping phases. *Bull. Seismol. Soc. Am.* **94**, 1762–1780.

Izutani, Y., Kanamori, H. (2001). Scale-dependence of seismic energy-to-moment ratio for strike-slip earthquakes in Japan. *Geophys. Res. Lett.* **28** (20), 4007–4010.

Kanamori, H., Rivera, L. (2004). Static and dynamic relations for earthquakes and their implications for rupture speed and stress drop. *Bull. Seismol. Soc. Am.* **94**, 314–319.

Kersten, R. (2005). Rockburst fractures—a unique example of fracture localization and propagation. In: Potvin, Y., Hudyma, M. (Eds.), Controlling Seismic Risk. Rockbursts and Seismicity in Mines. Australian Centre for Geomechanics, Nedlands, pp. 535–538.

Kijko, A., Lasocki, S., Graham, G., Retief, S.J.P. (2001). Non-parametric seismic hazard analysis in mines. In: van Aswegen, G., Durrheim, R.J., Ortlepp, W.D. (Eds.), Rockbursts and Seismicity in Mines. Dynamic Rock Mass Response to Mining. South African Inst. Min. Metal, Johannesburg, pp. 493–500.

Koch, K. (2002). Classification of local and regional events in central Europe based on estimates of S-wave spectral variance. *Geophys. J. Int.* **151**, 196–208.

Kwiatek, G. (2003). A search for the slow initial phase generated by seismic events at a copper mine in Poland. *Acta Geophys. Pol.* **51**, 369–385.

Kwiatek, G. (2004). A search for sequences of mining-induced seismic events at the Rudna copper mine in Poland. *Acta Geophys. Pol.* **52**, 155–171.

Lachenicht, R., Wiles, T., van Aswegen, G. (2001). Integration of deterministic modelling with seismic monitoring for the assessment of the rockmass response to mining: Part II Applications. In: van Aswegen, G., Durrheim, R.J., Ortlepp, W.D. (Eds.), Rockbursts and Seismicity in Mines. Dynamic Rock Mass Response to Mining. South African Inst. Min. Metal, Johannesburg, pp. 389–395.

Lasek, S., Matwiejszyn, A., Ptak, M. (2005). Borehole deformation preceding induced seismic events— Examples from coal mines in Upper Silesian Coal Basin in Poland. In: Potvin, Y., Hudyma, M. (Eds.), Controlling Seismic Risk. Rockbursts and Seismicity in Mines. Australian Centre for Geomechanics, Nedlands, pp. 485–491.

Lasocki, S., Kijko, A., Graham, G. (2002). Model-free seismic hazard analysis. In: Ogasawara, H., Yanagidani, T., Ando, M. (Eds.), Seismogenic Process Monitoring. Balkema, Lisse, pp. 327–339.

Li, S., Zheng, W., Trifu, C.I., Leslie, I. (2005). Implementation of a multi-channel microseismic monitoring system at Fankou lead–zinc mine, China. In: Potvin, Y., Hudyma, M. (Eds.), Controlling Seismic Risk. Rockbursts and Seismicity in Mines. Australian Centre for Geomechanics, Nedlands, pp. 613–616.

Li, Y., Doll Jr., C., Toksoz, M.N. (1995). Source characterization and fault plane determination for $m_{bLg} = 1.2$ to 4.4 earthquakes in the charlevoix seismic zone, Quebec, Canada. *Bull. Seismol. Soc. Am.* **85**, 1604–1621.

Li, Y., Thurber, C.H. (1988). Source properties of two microearthquakes at Kilauea Volcano, Hawaii. *Bull. Seismol. Soc. Am.* **78**, 1123–1132.

Luo, X., Jiang, F., Yang, S. (2005). A trial of microseismic monitoring of coal bumps at an underground coal mine in China. In: Potvin, Y., Hudyma, M. (Eds.), Controlling Seismic Risk. Rockbursts and Seismicity in Mines. Australian Centre for Geomechanics, Nedlands.

Madariaga, R. (1976). Dynamics of an expanding circular fault. *Bull. Seismol. Soc. Am.* **66**, 639–666.

Makowski, R. (2003). Source pulse estimation of mine shock by blind deconvolution. *Pure Appl. Geophys.* **160**, 1191–1205.

Malovichko, A.A. (2005). Study of "low-frequency" seismic events sources in the mines of the Verkhnekamskoye potash deposit. In: Potvin, Y., Hudyma, M. (Eds.), Controlling Seismic Risk. Rockbursts and Seismicity in Mines. Australian Centre for Geomechanics, Nedlands, pp. 373–377.

Malovichko, A.A., Baranov, Y.V. (2001). Application of numerical modelling for analysis of underground recordings of seismic events. In: van Aswegen, G., Durrheim, R.J., Ortlepp, W.D. (Eds.), Rockbursts and Seismicity in Mines. Dynamic Rock Mass Response to Mining. South African Inst. Min. Metal, Johannesburg, pp. 479–482.

Malovichko, A.A., Sabirov, R.H., Akhmetov, B.S. (2005). Ten years of seismic monitoring in mines of the Verkhnekamskoye potash deposit. In: Potvin, Y., Hudyma, M. (Eds.), Controlling Seismic Risk. Rockbursts and Seismicity in Mines. Australian Centre for Geomechanics, Nedlands, pp. 367–372.

Malovichko, A.A., Shulakov, D.Y., Dyaguilev, R.A., Sabirov, R.H., Ahmetov, B.S. (2001). Comprehensive monitoring of the large mine-collapse at the upper kama potash deposit in western ural. In: van Aswegen, G., Durrheim, R.J., Ortlepp, W.D. (Eds.), Rockbursts and Seismicity in Mines. Dynamic Rock Mass Response to Mining. South African Inst. Min. Metal, Johannesburg, pp. 309–312.

Mansurov, V.A. (2001). Prediction of rockbursts by analysis of induced seismicity data. *Int. J. Rock Mech. Min. Sci.* **38**, 893–901.

Marsan, D., Bean, C.J., Steacy, S., McCloske, J. (2000). Observation of diffusion processes in earthquake populations and implications for the predictability of seismicity systems. *J. Geophys. Res.* **105**, 28081–28094.

Marsan, D., Bean, C.J., Steacy, S., McCloskey, J. (1999). Spatio-temporal analysis of stress diffusion in a mining-induced seismicity system. *Geophys. Res. Lett.* **26**, 3697–3700.

Matwiejszyn, A., Ptak, M. (2002). Measurements of borehole deformations for assessment of rockburst hazard. In: Ogasawara, H., Yanagidani, T., Ando, M. (Eds.), Seismogenic Process Monitoring. Balkema, Lisse, pp. 63–74.

McGarr, A. (1992). Moment tensors of ten Witwatersrand mine tremors. *Pure Appl. Geophys.* **139**, 781–800.

McGarr, A. (1994). Some comparisons between mining-induced and laboratory earthquakes. *Pure Appl. Geophys.* **142**, 467–489.

McGarr, A. (1999). On relating apparent stress to the stress causing earthquake fault slip. *J. Geophys. Res.* **104**, 3003–3011.

McGarr, A. (2000). Energy budgets of mining-induced earthquakes and their interactions with nearby stopes. *Int. J. Rock Mech. Min. Sci.* **37**, 437–443.

McGarr, A. (2005). Observations concerning diverse mechanisms for mining-induced earthquakes. In: Potvin, Y., Hudyma, M. (Eds.), Controlling Seismic Risk. Rockbursts and Seismicity in Mines. Australian Centre for Geomechanics, Nedlands, pp. 107–111.

McGarr, A., Fletcher, J.B. (2003). Maximum slip in earthquake fault zones, apparent stress, and stick-slip friction. *Bull. Seismol. Soc. Am.* **93**, 2355–2362.

McGarr, A., Fletcher, J.B. (2005). Development of ground motion prediction equations relevant to shallow mining-induced seismicity in the Trail Mountain area, Emery County, Utah. *Bull. Seismol. Soc. Am.* **95**, 31–47, doi:10.1785/0120040046.

McGill, R.B. (2005). Strategies to manage seismicity in a deep VCR environment on Mponeng. In: Potvin, Y., Hudyma, M. (Eds.), Controlling Seismic Risk. Rockbursts and Seismicity in Mines. Australian Centre for Geomechanics, Nedlands, pp. 419–424.

Mendecki, A.J. (1993). Real time quantitative seismology in mines. In: Young, R.P. (Ed.), Rockbursts and Seismicity in Mines. Balkema, Rottterdam, pp. 287–295.

Mendecki, A.J. (2005). Persistence of seismic rockmass response to mining. In: Potvin, Y., Hudyma, M. (Eds.), Controlling Seismic Risk. Rockbursts and Seismicity in Mines. Australian Centre for Geomechanics, Nedlands.

Mercer, R.A., Bawden, W.F. (2005a). A statistical approach for the integrated analysis of mine-induced seismicity and numerical stress estimates, a case study—Part I: Developing the relations. *Int. J. Rock Mech. Min. Sci.* **42**, 47–72.

Mercer, R.A., Bawden, W.F. (2005b). A statistical approach for the integrated analysis of mine-induced seismicity and numerical stress estimates, a case study—Part II: Evaluation of the relations. *Int. J. Rock Mech. Min. Sci.* **42**, 73–94.

Milev, A.M., Spottiswoode, S.M. (2002). Effect of the rock properties on mining-induced seismicity around the Ventersdorp Contact Reef, Witwatersrand, South Africa. *Pure Appl. Geophys.* **159**, 165–177.

Mortimer, Z. (2002). Studies on the chaotic dynamics of mining induced seismicity. In: Ogasawara, H., Yanagidani, T., Ando, M. (Eds.), Seismogenic Process Monitoring. Balkema, Lisse, pp. 341–354.

Mueller, C.S. (1985). Source pulse enhancement by deconvolution of an empirical Green's function. *Geophys. Res. Lett.* **12** (1), 33–36.

Nagai, N., Ando, M., Ogasawara, H., Okhura, T., Iio, Y., Cho, A. (2002). Location and temporal variations of shear wave splitting in a South African gold mine. In: Ogasawara, H., Yanagidani, T., Ando, M. (Eds.), Seismogenic Process Monitoring. Balkema, Lisse, pp. 185–198.

Naoi, M., Ogasawara, H., Takeuchi, J., Yamamoto, A., Shimoda, N., Morishita, K., Ishii, H., Nakao, S., van Aswegen, G., Mendecki, A.J., Lenegan, P., Ebrahim-Trollope, R., Iio, Y. (2006). Small slow-strain steps and their forerunners observed in gold mine in South Africa. *Geophys. Res. Lett.* **33**, L12304, doi:10.1029/2006GL026507.

Ogasawara, H., Sato, S., Nishii, S., Sumitomo, N., Nakao, H.I.Y.I.S., Ando, M., Takano, K., Nagai, N., Ohkura, T., Kawakata, H., Satoh, T., Kusunose, K., Cho, A., Mendecki, A.J., Cichowicz, A., Green, R., Kataka, M.O. (2001). Semi-controlled seismogenic experiments in south african deep gold mines. In: van Aswegen, G., Durrheim, R.J., Ortlepp, W.D. (Eds.), Rockbursts and Seismicity in Mines. Dynamic Rock Mass Response to Mining. South African Inst. Min. Metal, Johannesburg, pp. 293–300.

Ogasawara, H., Yanagidani, T., Ando, M. (Eds.) (2002a). Seismogenic Process Monitoring. Balkema, Lisse.

Ogasawara, H., and the research group for semi-controlled earthquake-generation experiments in South African deep gold mines (2002b). Review of semi-controlled earthquake-generation experiments in South African deep gold mines (1992–2001). In: Ogasawara, H., Yanagidani, T., Ando, M. (Eds.), Seismogenic Process Monitoring. Balkema, Lisse, pp. 119–150.

Ogasawara, H., Miwa, T., and the research group for semi-controlled earthquake-generation experiments in South African deep gold mines (2002c). Microearthquake scaling relationship using near-source, redundant, wide-dynamic-range accelerograms in a South African deep gold mine. In: Ogasawara, H., Yanagidani, T., Ando, M. (Eds.), Seismogenic Process Monitoring. Balkema, Lisse, pp. 151–164.

Ogasawara, H., Takeuchi, J., Shimoda, N., Ishii, H., Nakao, S., van Aswegen, G., Mendecki, A.J., Cichowicz, A., Ebrahim-Trollope, R., Kawakata, H., Iio, Y., Ohkura, T., Ando, M. (2005a). High-resolution strain monitoring during $m > 2$ events in a south african deep gold mine in close proximity to hypocentres. In: Potvin, Y., Hudyma, M. (Eds.), Controlling Seismic Risk. Rockbursts and Seismicity in Mines. Australian Centre for Geomechanics, Nedlands, pp. 385–391.

Ogasawara, H., Takeuchi, J., Shimoda, N., Nakatani, M., Kato, M., Iio, Y., Kawakata, H., Yamada, T., Yamauchi, T., Ishii, H., Sato, T., Kusunose, K., Otsuki, K., Kita, S., Nakao, S., Ward, A.K., McGill, R.B., Murphy, S.K., Mendecki, A., Aswegen, G. (2005b). Multidisciplinary monitoring of the entire life span of an earthquake in south african gold mines. In: Potvin, Y., Hudyma, M. (Eds.), Controlling Seismic Risk. Rockbursts and Seismicity in Mines. Australian Centre for Geomechanics, Nedlands, pp. 393–398.

Ortlepp, W.D. (1992). Note on fault-slip motion inferred from a study of micro-cataclastic particles from an underground shear rupture. *Pure Appl. Geophys.* **139**, 677–695.

Ortlepp, W.D. (2000). Observation of mining-induced faults in an intact rock mass at depth. *Int. J. Rock Mech. Min. Sci.* **37**, 423–436.

Ortlepp, W.D. (2001). Thoughts on the rockburst source mechanism based on observations of the mine-induced shear rupture. In: van Aswegen, G., Durrheim, R.J., Ortlepp, W.D. (Eds.), Rockbursts and Seismicity in Mines. Dynamic Rock Mass Response to Mining. South African Inst. Min. Metal, Johannesburg, pp. 43–51.

Ortlepp, W.D. (2005). RaSiM comes of age—a review of the contribution to the understanding and control of mine rockbursts. In: Potvin, Y., Hudyma, M. (Eds.), Controlling Seismic Risk. Rockbursts and Seismicity in Mines. Australian Centre for Geomechanics, Nedlands, pp. 3–20.

Ortlepp, W.D., Armstrong, R., Ryder, J.A., O'Connor, D. (2005). Fundamental study of micro-fracturing on the slip surface of mine-induced dynamic brittle shear zones. In: Potvin, Y., Hudyma, M. (Eds.), Controlling Seismic Risk. Rockbursts and Seismicity in Mines. Australian Centre for Geomechanics, Nedlands, pp. 229–237.

Ouillon, G., Sornette, D. (2000). The concept of 'critical earthquakes' applied to mine rockbursts with time-to-failure analysis. *Geophys. J. Int.* **143**, 454–468.

Oye, V., Bungum, H., Roth, M. (2005). Source parameters and scaling relations for mining-related seismicity within the Pyhasalmi ore mine, Finland. *Bull. Seismol. Soc. Am.* **95**, 1011–1026, doi:10.1785/0120040170.

Panza, G.F., Sarao, A. (2000). Monitoring volcanic and geothermal areas by full seismic moment tensor inversion: are non-double-couple components always artefacts of modelling? *Geophys. J. Int.* **143**, 353–364.

Piana, M., Bertero, M. (1997). Projected Landweber method and preconditioning. *Inv. Prob.* **13**, 441–463.

Potvin, Y., Hudyma, M. (Eds.) (2005). Controlling Seismic Risk. Rockbursts and Seismicity in Mines. Australian Centre for Geomechanics, Nedlands.

Potvin, Y., Hudyma, M.R. (2001). Seismic monitoring in highly mechanized hardrock mines in Canada and Australia. In: van Aswegen, G., Durrheim, R.J., Ortlepp, W.D. (Eds.), Rockbursts and Seismicity in Mines. Dynamic Rock Mass Response to Mining. South African Inst. Min. Metal, Johannesburg, pp. 267–280.

Prejean, S.G., Ellsworth, W.L. (2001). Observations of earthquake source parameters at 2 km depth in the Long Valley caldera, eastern California. *Bull. Seismol. Soc. Am.* **91**, 165–177.

Pytel, W. (2003). Rock mass—mine workings interaction model for Polish copper mine conditions. *Int. J. Rock Mech. Min. Sci.* **40**, 497–526.

Reyes-Montes, J.M., Rietbrock, A., Collins, D.S., Young, R.P. (2005). Relative location of excavation induced microseismicity at the Underground Research Laboratory (AECL, Canada) using surveyed reference events. *Geophys. Res. Lett.* **32**, L05308, doi:10.1029/2004GL021733.

Richardson, E., Jordan, T.H. (2001). Some properties of gold-mine seismicity and implications for tectonic earthquakes. In: van Aswegen, G., Durrheim, R.J., Ortlepp, W.D. (Eds.), Rockbursts and Seismicity in Mines. Dynamic Rock Mass Response to Mining. South African Inst. Min. Metal, Johannesburg, pp. 149–156.

Richardson, E., Jordan, T.H. (2002). Seismicity in deep gold mines of South Africa: Implications for tectonic earthquakes. *Bull. Seismol. Soc. Am.* **92**, 1766–1782.

Richardson, E., Nyblade, A.A., Walter, W.R., Rodgers, A.J. (2005). Source characteristics of mining-induced seismicity from moment-tensor analysis and spatio-temporal relationships. In: Potvin, Y., Hudyma, M. (Eds.), Controlling Seismic Risk. Rockbursts and Seismicity in Mines. Australian Centre for Geomechanics, Nedlands, pp. 123–127.

Riemer, K.L. (2005). Interpreting complex waveforms from some mining related seismic events. In: Potvin, Y., Hudyma, M. (Eds.), Controlling Seismic Risk. Rockbursts and Seismicity in Mines. Australian Centre for Geomechanics, Nedlands.

Roth, M., Bungum, H. (2003). Waveform modeling of the 17 August 1999 Kola peninsula earthquake. *Bull. Seismol. Soc. Am.* **93**, 1559–1572.

Rudziński, L., Dębski, W. (2008). Mining events relocation with the double-difference method. *Acta Geodyn. Geomater.* **5** (2), 1–8.

Satoh, T., and the research group for semi-controlled earthquake-generation experiments in South African deep gold mines (2002). Near source observation of small initial phase generated by earthquakes in a deep gold mine in South Africa. In: Ogasawara, H., Yanagidani, T., Ando, M. (Eds.), Seismogenic Process Monitoring. Balkema, Lisse, pp. 166–171.

Savage, J.C., Wood, M.D. (1971). The relation between apparent stress and stress drop. *Bull. Seismol. Soc. Am.* **61**, 1381–1388.

Sellers, E.J., Kataka, M.O., Linzer, L.M. (2003). Source parameters of acoustic emission events and scaling with mining-induced seismicity. *J. Geophys. Res.* **108**, 2418, doi:10.1029/2001JB000670.

Senatorski, P. (2001). Slip weakening and energy–moment scaling relationship. *Acta Geophys. Pol.* **49**, 75–93.

Senatorski, P. (2007). Apparent stress scaling and statistical trends. *Phys. Earth Planet. Int.* **160**, 230–244.

Senfaute, G., Al-Heib, M., Josein, J.P., Noirel, J.F. (2001). Detection and monitoring of high stress concentration zones induced by coal mining using numerical and microseismic methods. In: van Aswegen, G., Durrheim, R.J., Ortlepp, W.D. (Eds.), Rockbursts and Seismicity in Mines. Dynamic Rock Mass Response to Mining. South African Inst. Min. Metal, Johannesburg, pp. 453–456.

Šileny, J., Milev, A. (2005). Source mechanism—Dipole versus single force application to mining induced seismic events in deep level gold mines in South Africa. In: Potvin, Y., Hudyma, M. (Eds.), Controlling Seismic Risk. Rockbursts and Seismicity in Mines. Australian Centre for Geomechanics, Nedlands, pp. 259–265.

Šileny, J., Milev, A. (2006). Seismic moment tensor resolution on a local scale: Simulated rockburst and mine-induced seismic event in the Kopanga gold mine, South Africa. *Pure Appl. Geophys.* **163**, 1495–1513.

Šileny, J., Pšenčík, I., Young, R.P. (2001). Point-source inversion neglecting a nearby free surface: simulation of the Underground Research Laboratory, Canada. *Geophys. J. Int.* **146**, 171–180.

Šileny, J., Vavryčuk, V. (2002). Can unbiased source be retrieved from anisotropic waveforms by using an isotropic model of the medium? *Tectonophysics* **356**, 125–138.

Sivakumar, C., Srinivasan, C., Gupta, R.N. (2005). Microseismic monitoring to study strata behaviour and to predict rockfalls on a longwall face at Rajendra Colliery in India. In: Potvin, Y., Hudyma, M. (Eds.), Controlling Seismic Risk. Rockbursts and Seismicity in Mines. Australian Centre for Geomechanics, Nedlands, pp. 135–144.

Sornette, D. (2001). The concept of 'critical rupture' and 'finite-time singularity' applied to mine rockbursts: A review. In: van Aswegen, G., Durrheim, R.J., Ortlepp, W.D. (Eds.), Rockbursts and Seismicity in Mines. Dynamic Rock Mass Response to Mining. South African Inst. Min. Metal, Johannesburg, pp. 483–488.

Spottiswoode, S.M. (2001). Synthetic seismicity mimics observed seismicity in deep tabular mines. In: van Aswegen, G., Durrheim, R.J., Ortlepp, W.D. (Eds.), Rockbursts and Seismicity in Mines. Dynamic Rock Mass Response to Mining. South African Inst. Min. Metal, Johannesburg, pp. 371–377.

Spottiswoode, S.M. (2005). Mine layout design and medium-term prediction of seismicity. In: Potvin, Y., Hudyma, M. (Eds.), Controlling Seismic Risk. Rockbursts and Seismicity in Mines. Australian Centre for Geomechanics, Nedlands, pp. 211–217.

Sprenke, K.F., White, B.G., Rohay, A.C., Whyatt, J.K., Stickney, M.C. (2002). Comparison of body-wave displacement with damage observations of a rockburst, Coeur d'Alene Mining District, Idaho. *Bull. Seismol. Soc. Am.* **92**, 3321–3328.

Srinivasan, C., Arora, S.K., Benady, S. (1999). Precursory monitoring of impending rockbursts in kolar gold mines from microseismic emissions at deeper levels. *Int. J. Rock Mech. Min. Sci* **36**, 941–948.

Stec, K. (2007). Characteristics of seismic activity of the Upper Silesian Coal Basin in Poland. *Geophys. J. Int.* **168**, 757–768.

Stewart, R.A., Reimold, W.U., Charlesworth, E.G., Ortlepp, W.D. (2001). The nature of a deformation zone and fault rock related to a recent rockburst at Western Deep Levels Gold Mine, Witwatersrand Basin, South Africa. *Tectonophysics* **337**, 173–190.

Stork, A.L., Ito, H. (2004). Source parameter scaling for small earthquakes observed at the western nagano 800 m-deep borehole, central Japan. *Bull. Seismol. Soc. Am.* **94**, 1781–1796.

Swanson, P. (2001). Development of an automated pc-network-based seismic monitoring system. In: van Aswegen, G., Durrheim, R.J., Ortlepp, W.D. (Eds.), Rockbursts and Seismicity in Mines. Dynamic Rock Mass Response to Mining. South African Inst. Min. Metal, Johannesburg, pp. 11–17.

Talebi, S., Côté, M. (2005). Implosional focal mechanisms in a hard-rock mine. In: Potvin, Y., Hudyma, M. (Eds.), Controlling Seismic Risk. Rockbursts and Seismicity in Mines. Australian Centre for Geomechanics, Nedlands, pp. 113–121.

Teyssoneyre, V., Feignier, B., Šileny, J., Coutant, O. (2002). Moment tensor inversion of regional phases: Application to a mine collapse. *Pure Appl. Geophys.* **159**, 111–130.
Trifu, C.-I. (2001). Assessing the rockmass condition by the analysis of failure mechanisms. In: van Aswegen, G., Durrheim, R.J., Ortlepp, W.D. (Eds.), Rockbursts and Seismicity in Mines. Dynamic Rock Mass Response to Mining. South African Inst. Min. Metal, Johannesburg, pp. 75–80.
Trifu, C.-I. (Ed.) (2002). The Mechanism of Induced Seismicity. Birkhauser Verlag, Basel.
Trifu, C.-I., Angus, D., Shumila, V. (2000). A fast evaluation of the seismic moment tensor for induced seismicity. *Bull. Seismol. Soc. Am.* **90**, 1521–1527.
Trifu, C.-I., Shumila, V. (2002a). Reliability of seismic moment tensor inversions for induced microseismicity at Kidd mine, Ontario. *Pure Appl. Geophys.* **159**, 145–164.
Trifu, C.-I., Shumila, V. (2002b). The use of uniaxial recordings in moment tensor inversions for induced seismic sources. *Tectonophysics* **356**, 171–180.
Urbancic, T.I., Trifu, C.I. (1996). Effects of rupture complexity and stress regime on scaling relations of induced microseismic events. *Pure Appl. Geophys.* **147**, 319–343.
Urbancic, T.-I., Trifu, C.I. (2000). Recent advances in seismic monitoring technology at Canadian mines. *J. Appl. Geophys.* **45**, 225–237.
Urbancic, T.I., Young, R.P. (1993). Space-time variations in source parameters of mining-induced seismic events with $m < 0$. *Bull. Seismol. Soc. Am.* **83**, 378–397.
van Aswegen, G., Durrheim, R.J., Ortlepp, W.D. (Eds.) (2001). Rockbursts and Seismicity in Mines. Dynamic Rock Mass Response to Mining. South African Inst. Min. Metal, Johannesburg.
Venkataraman, A., Beroza, G.C., Ide, S., Imanishi, K., Ito, H., Iio, Y. (2006). Measurements of spectral similarity for microearthquakes in western Nagano, Japan. *J. Geophys. Res.* **111**, B03303, doi:10.1029/2005JB003834.
Venkataraman, A., Boatwright, J., Beroza, G.C. (2006). A brief review of techniques used to estimate radiated seismic energy. In: Abercrombie, R., McGarr, A., Kanamori, H., Di Toro, G. (Eds.), Earthquakes: Radiation Energy and the Physics of Faulting. In: Geophys. Monogr. Ser, vol. 170, AGU, Washington, pp. 15–24.
Waldhauser, F., Ellsworth, W. (2000). A double-difference earthquake location algorithm: method and application to the northern Hayward fault. *Bull. Seismol. Soc. Am.* **90**, 1353–1368.
Wiles, T. (2005). Rockburst prediction using numerical modelling—realistic limits for failure prediction accuracy. In: Potvin, Y., Hudyma, M. (Eds.), Controlling Seismic Risk. Rockbursts and Seismicity in Mines. Australian Centre for Geomechanics, Nedlands, pp. 57–63.
Wiles, T., Lachenicht, R., van Aswegen, G. (2001). Integration of deterministic modelling with seismic monitoring for the assessment of the rockmass response to mining: Part 1 Theory. In: van Aswegen, G., Durrheim, R.J., Ortlepp, W.D. (Eds.), Rockbursts and Seismicity in Mines. Dynamic Rock Mass Response to Mining. South African Inst. Min. Metal, Johannesburg, pp. 379–387.
Wyss, M., Brune, J. (1968). Seismic moment, stress, and source dimension for earthquakes in the California-Nevada region. *J. Geophys. Res.* **73**, 4681–4694.
Yamada, T., Mori, J.J., Ide, S., Abercrombie, R.E., Kawakata, H., Nakatani, M., Iio, Y., Ogasawara, H. (2007). Stress drops and radiated seismic energy of microearthquakes in a south african gold mine. *J. Geophys. Res.* **112**, B03305, doi:10.1029/2006JB004553.
Yamada, T., Mori, J.J., Ide, S., Kawakata, H., Iio, Y., Ogasawara, H. (2005a). Correction to radiation efficiency and apparent stress of small earthquakes in a south african gold mine. *J. Geophys. Res.* **110**, B06301, doi:10.1029/2005JB003789.
Yamada, T., Mori, J.J., Ide, S., Kawakata, H., Iio, Y., Ogasawara, H. (2005b). Radiation efficiency and apparent stress of small earthquakes in a South African gold mine. *J. Geophys. Res.* **110**, B01305, doi:10.1029/2004JB002221.
Young, R.P., Collins, D.S. (2001). Seismic study of rock fracture at the Underground Research Laboratory, Canada. *Int. J. Rock Mech. Min. Sci.* **38**, 787–799.
Young, R.P., Collins, D.S., Reyes-Montes, J.M., Baker, C. (2004). Quantification and interpretation of seismicity. *Int. J. Rock Mech. Min. Sci.* **41**, 1317–1327.

INDUCED SEISMICITY IN HYDROCARBON FIELDS

JENNY SUCKALE

ABSTRACT

Over the last few decades, it has become clear that various human activities have the potential to generate seismic activity. Examples include subsurface waste injection, reservoir impoundment in the vicinity of large dams, and development of mining, geothermal or hydrocarbon resources. Recently, induced seismicity has also become a concern in connection with geologic carbon sequestration projects. This study focuses on seismicity induced by hydrocarbon production by summarizing the published case studies and describing current theoretical approaches to model these. It is important to understand the conditions under which hydrocarbon production may lead to seismic activity in order to ensure that they are performed safely. Our knowledge of induced seismicity in hydrocarbon fields has progressed substantially over the last few years owing to more intensive high-quality instrumentation of oil fields and a continuous effort to understand the phenomenon theoretically. However, much of the available literature is dispersed over a variety of journals and specialized reports. This review aims at providing a first step toward making the current knowledge about induced seismicity in hydrocarbon fields more accessible to a broad audience of scientists.

KEY WORDS: Induced seismicity, hydrocarbon fields, poroelasticity, seismic hazard, monitoring, fluid extraction, fluid injection, midcrustal earthquakes, microseismicity.   © 2009 Elsevier Inc.

1. INTRODUCTION

Numerous human activities are known to induce seismic activity of varying magnitudes. Patterns of induced seismicity vary from (1) continued microseismicity (e.g. mines as well as hydrocarbon and geothermal fields), (2) intense microseismicity for a short time after a certain event (e.g. hydraulic fracturing), (3) moderate seismic activity in the magnitude range between two and four (e.g. hydrocarbon fields and mines), to (4) catastrophic activity of magnitude 6 and above (e.g. reservoir induced seismicity). The largest induced earthquakes have been related to reservoir impoundments at Koyna Dam, India (Guha *et al.*, 1971; Singh *et al.*, 1975; Gupta and Rastogi, 1976), Hsingfengchian (also Xinfengjiang) Dam, China (Shen *et al.*, 1974; Miaoyueh *et al.*, 1976; Wang *et al.*, 1976), Kariba Dam, Zambia/Zimbabwe (Gough and Gough, 1970), and Kremasta, Greece (Simpson, 1976; Stein *et al.*, 1982). At least two of these, Kremasta and Koyna, caused deaths, injury and extensive property damage (Simpson, 1976; Guha, 2000).

Induced seismicity in hydrocarbon fields is typically small to moderate ($M_L \leq 4.5$). However, a connection to hydrocarbon production has also been suspected for two destructive earthquake sequences at Coalinga, USA, 1983-1987, and Gazli, Uzbekistan, 1976-1984 (see Section 6). While available observational evidence and theoretical modeling indicate that the Coalinga sequence was probably not induced, the case of the earthquakes at Gazli field remains controversial. Should there be a connection between the Gazli earthquakes and hydrocarbon production, then these events would constitute the largest seismic events caused by human activity until today.

Before delving into a discussion of various cases for which a connection between hydrocarbon production and seismic activity might exist, it is worth noting that which events are considered to be "induced" is largely subjective. Quantitative models (Sections 4 and 5) focus on assessing changes in stress and/or pore pressure, but it is unlikely that any event is driven by these perturbing effects alone. The pre-existing stress field also plays an important role. Thus, a key challenge lies in deciding whether a given, production-related perturbation in stress or pore pressure is significant. This challenge is usually further exacerbated by lack of knowledge of the pre-existing stress field and exact production levels.

We have assembled 70 cases of hydrocarbon fields for which a connection between unusual seismicity patterns and hydrocarbon production has been suggested in the scientific literature (see Table 1). A striking feature is that occurrence of induced seismicity seems to have a distinctly regional character. It is most commonly observed only in two sedimentary basins, namely in the Permian basin, Texas, and in the Rotliegendes, Netherlands. In both regions, there were attempts to identify criteria why seismicity is induced only in certain fields and not in others, but with limited success (Doser *et al.*, 1992; Van Eijs *et al.*, 2006).

Not all induced seismicity in hydrocarbon fields is unwanted: hydraulic fracturing is commonly used to increase the permeability of a hydrocarbon reservoir through intentional creation of new fractures. Monitoring of the associated microseismicity can give important insight into the fracture pattern and fluid migration. Despite its importance for hydrocarbon development, we largely refrain from discussing hydraulic fracturing in this review. The reason is that during hydraulic fracturing seismicity is induced intentionally and thus constitutes a different phenomenon meriting separate treatment. That being said, it is inevitable to touch on hydraulic fracturing briefly in the context of seismic-monitoring techniques in hydrocarbon fields, since these draw heavily on advances made in the context of hydraulic fracturing.

The main focus of this review is a compilation of the observational evidence for induced seismicity in hydrocarbon fields and theoretical attempts to understand it (for a previous review see Grasso (1992b)). Induced seismicity in hydrocarbon fields can be broadly classified as being primarily related either to fluid injection or to fluid extraction. Needless to say, most cases cannot be easily grouped into one of these categories, because of concurrent fluid extraction (reservoir depletion) and injection (secondary recovery) and because fracturing of the involved geological units can lead to complex interconnected flow patterns throughout the field. Nevertheless, this categorization has proven useful in thinking about induced seismicity and it provides the starting point for most theoretical models.

This review is structured in the following way. In Section 2, we list all hydrocarbon fields for which a connection between seismicity and hydrocarbon production has been suggested in the scientific literature. We also summarize the most common observations regarding induced seismicity in hydrocarbon fields and briefly discuss the phenomenon that cases of induced seismicity are particularly common in certain regions.

Section 3 briefly discusses recent advances in seismic monitoring of hydrocarbon fields experiencing induced seismicity. Section 4 deals with injection induced seismicity and Section 5 with seismicity induced by the extraction of fluids. Section 6 reviews the controversy whether hydrocarbon production could induce major midcrustal earthquakes. Finally, Section 7 gives a brief summary and highlights important open questions.

2. Case Studies, Common Observations and Regional Occurrence

2.1. Documented Case Studies

The total number of hydrocarbon fields in which induced seismicity has been reported is difficult to pinpoint. An overview of all cases for which induced seismicity has been suspected in the scientific literature is given in Table 1. It is important to keep in mind that this list should not imply that all of these hydrocarbon fields truly exhibit induced seismicity. For some cases, a connection is generally agreed upon (e.g. Lacq, France), while for others (in particular: Coalinga, Kettleman, and Montebello fields, California) it is controversial or even unlikely. Contrary to a prior compilation (Grasso, 1992b), we list all cases discussed in the literature and not only those in which seismic activity exceeded a certain minimum magnitude. The reason is that information about magnitudes is not always available and, if it is, it might not be comparable for different fields due to the utilization of different magnitude scales. Finally, we stress that Table 1 does not contain hydrocarbon fields in which microseismicity was induced intentionally through hydraulic fracturing experiments.

The overview in Table 1 is inevitably incomplete for several reasons. (1) Induced seismicity in hydrocarbon fields typically falls into a small to moderate magnitude range. Thus, it might not be detected unless a local network is operated in its immediate vicinity. (2) The presence of natural seismicity can obscure the existence of induced events. (3) A lot of research regarding induced seismicity is done by the industry and not always publicly available.

2.2. Common Observations

Induced seismicity is related to stress changes in the reservoir and surrounding rocks that can be caused by various mechanisms ranging from pore pressure variations, to geochemical reactions, temperature effects, and either locking or reactivation of pre-existing faults. Accordingly, the patterns of induced seismicity vary a lot for different fields or events within the same field over space and time. For example, the substantial subsidence at Ekofisk field, North Sea, Norway, was considered as largely aseismic. Nonetheless, a sizeable event $M_w = 4.1 - 4.4$ occurred within the field in May 2002 (Ottemöller et al., 2005). A map of the regional seismicity around Ekofisk field is shown in Figs 1 and 2. Despite this variability in seismicity patterns, a few general observations can be made.

1. *Magnitude Range:* In most hydrocarbon fields, induced seismic activity has been limited to small and moderate magnitudes ($M_L \leq 4.5$).
2. *Correlation with production:* It is often challenging to determine the correlation between seismicity and hydrocarbon production exactly because of the lack of data detailing the production pattern at various wells. In cases where production data is

TABLE 1. Documented cases of induced seismicity in hydrocarbon fields

Hydrocarbon field	Country	Citation	Potential cause
Snipe Lake	Alberta, Canada	Milne (1970),	
Strachan	Alberta, Canada	Wetmiller (1986),	Extraction
		Baranova et al. (1999)	
Gobles	Appalachian, Canada	Mereu et al. (1986)	Injection
Eagle&Eagle West	British Columbia, Canada	Horner et al. (1994)	Extraction
Shengli	Shandong Province, China	Shouzhong et al. (1987)	Injection
Dan	North Sea, Denmark	Ovens et al. (1997)	Extraction
Lacq	Aquitaine, France	Grasso and Wittlinger (1990), add. references in Section 5	
Meillon	Aquitaine, France	Grasso (1992a)	Extraction
Söhlingen/Rotenburg	Rotliegendes, Germany	Dahm et al. (2007)	
Caviaga	Po Valley, Italy	Calloi et al. (1956)	
Umm Gudair	Kuwait	Bou-Rabee and Nur (2002)	Burning
Burgan	Kuwait	Bou-Rabee and Nur (2002)	Burning
Assen	Rotliegendes, Netherlands	Haak (1991)	
Groningen	Rotliegendes, Netherlands	Gussinklo et al. (2001), Van Wees et al. (2003), Van Eijs et al. (2006)	Fault reactivation?
Roswinkel	Rotliegendes, Netherlands	Van Eijs et al. (2006)	
Bergermeer	Rotliegendes, Netherlands	Van Eijs et al. (2006)	
Eleveld	Rotliegendes, Netherlands	Van Eijs et al. (2006), Roest and Kuilman (1994)	Fault reactivation
Bergen	Rotliegendes, Netherlands	Van Eijs et al. (2006)	

Table 1 Continued.

Hydrocarbon field	Country	Citation	Potential cause
Annerveen	Rotliegendes, Netherlands	Van Eijs et al. (2006), Mulders (2003)	
Appelscha	Rotliegendes, Netherlands	Van Eijs et al. (2006)	
Emmen	Rotliegendes, Netherlands	Van Eijs et al. (2006)	
Dalen	Rotliegendes, Netherlands	Van Eijs et al. (2006)	
Roden	Rotliegendes, Netherlands	Van Eijs et al. (2006)	
VriesNoord	Rotliegendes, Netherlands	Van Eijs et al. (2006)	
Ureterp	Rotliegendes, Netherlands	Van Eijs et al. (2006)	
Emmen-Nw. A'Dam	Rotliegendes, Netherlands	Van Eijs et al. (2006)	
Schoonebeek	Rotliegendes, Netherlands	Van Eijs et al. (2006)	
VriesCentraal	Rotliegendes, Netherlands	Van Eijs et al. (2006)	
Coevorden	Rotliegendes, Netherlands	Van Eijs et al. (2006)	
Ekofisk	North Sea, Norway	Zoback and Zinke (2002), Ottemöller et al. (2005)	Various
Valhall	North Sea, Norway	Zoback and Zinke (2002)	Extraction/fault reactivation
Visund	North Sea, Norway	Wiprut and Zoback (2000)	Various
Shuaiba	Oman	Sze (2005)	
Romashkino	Volga-Ural, Russia	Adushkin et al. (2000), Turuntaev and Razumnaya (2002)	
Novo-Elkhovskoye	Volga-Ural, Russia	Adushkin et al. (2000)	
Starogroznenskoye	Russia	Kouznetsov et al. (1994)	
Barsa-Gelmes-Vishka	Russia	Kouznetsov et al. (1994)	

(continued on next page)

Table 1 Continued.

Hydrocarbon field	Country	Citation	Potential cause
Gudermes	North Caucasus, Russia	Smirnova (1968)	Fault reactivation
Grozny	Tchetchny	Grasso (1992b) and references therein	Extraction (?)
Big Escambia Creek	Alabama, USA	Gomberg and Wolf (1999)	Extraction (?)
Little Rock	Alabama, USA	Gomberg and Wolf (1999)	Extraction (?)
Sizemore Creek	Alabama, USA	Gomberg and Wolf (1999)	Extraction (?)
Coalinga	California, USA	McGarr (1991)	Isostasy (?)
Kettleman	California, USA	McGarr (1991)	Isostasy (?)
Montebello	California, USA	McGarr (1991)	Isostasy (?)
Orcutt	California, USA	Kanamori and Hauksson (1992)	Injection
Wilmington	California, USA	Kovach (1974)	Extraction
Rangely	Colorado, USA	Raleigh et al. (1976)	Experiment
South Eugene Island	Louisiana, USA	Chan et al. (2002)	Injection
Hunt	Mississippi, USA	Nicholson and Wesson (1990)	Injection (?)
Sleepy Hollow	Nebraska, USA	Rothe and Lui (1983), Evans and Steeples (1987)	Injection (?)
Catoosa District	Oklahoma, USA	Nicholson and Wesson (1990)	Injection?
Love County	Oklahoma, USA	Nicholson and Wesson (1990)	Injection
East Durant	Oklahoma, USA	Nicholson and Wesson (1990)	
Gobles	Ontario, USA	Mereu et al. (1986), Nicholson and Wesson (1990)	Injection
Apollo-Hendrick	Texas, USA	Doser et al. (1992)	Injection (?)
Blue Ridge	Texas, USA	Kreitler (1978)	Extraction/subsidence

Table 1 Continued.

Hydrocarbon field	Country	Citation	Potential cause
Clinton	Texas, USA	Kreitler (1978), Sheets (1979)	Extraction/subsidence
Cogdell Canyon Reef	Texas, USA	Davis and Pennington (1989)	Injection (?)
Fashing	Texas, USA	Pennington et al. (1986), Davis et al. (1995)	Extraction/fault reactivation
Goose Creek	Texas, USA	Pratt and Johnson (1926), Kreitler (1978), Sheets (1979)	Extraction/subsidence
Imogene	Texas, USA	Pennington et al. (1986)	Extraction/fault reactivation
Kermit	Texas, USA	Nicholson and Wesson (1990)	
Keystone	Texas, USA	Marroouin et al. (1992), Orr and Keller (1981)	Injection
Mykawa	Texas, USA	Yerkes and Castle (1976), Kreitler (1978)	Extraction/subsidence
South Houston	Texas, USA	Kreitler (1978)	Extraction/subsidence
War-Wink	Texas, USA	Baker et al. (1989), Doser et al. (1991), Baker et al. (1991)	Extraction (?)
Webster	Texas, USA	Kreitler (1978)	Extraction/subsidence
Dollarhide	Texas/New Mexico, USA	Nicholson and Wesson (1990)	Injection (?)
Gazli	Uzbekistan	Simpson and Leith (1985)	Variable production rates (?)
Costa Oriental	Lake Maracaibo, Venezuela	Murria (1997)	Extraction/subsidence

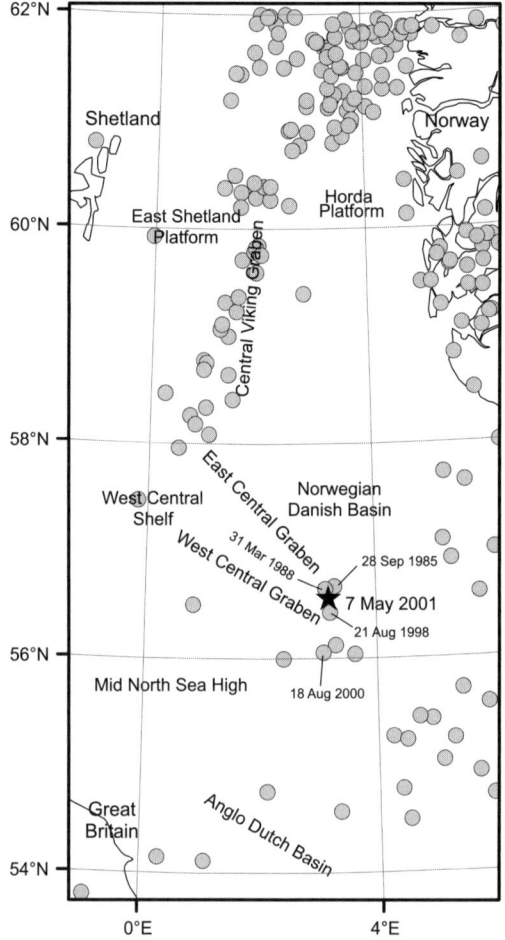

FIG. 1. Map of the regional seismicity during the time period 1970-2003 recorded in the vicinity of Ekofisk field. The seismic data was merged from University of Bergen, British Geological Survey and the International Seismological Centre catalogs. Relevant tectonic features in the area are labeled. The epicenters are indicated by circles independent of their magnitudes and the event on 7 May 2001 studied by Ottemöller *et al.* (2005) is marked by a star. Previously to this event only four notable events labeled by their date of occurrence were recorded for Ekofisk field. The outline of the Ekofisk and neighboring fields in comparison to the event on 7 May 2001 is shown in Fig. 1. *Source:* Figure reproduced from Ottemöller et al. (2005, Fig. 1, p. 2).

available, an obvious correlation between production and seismic activity – in the sense of elevated seismicity levels immediately after increases in production rate – is rarely observed. An important exception is the controlled experiment at Rangely field (see Section 4.1). Furthermore, more complex correlations between temporal and spatial patterns of seismicity and production have been reported for various cases (see Sections 4 and 5).

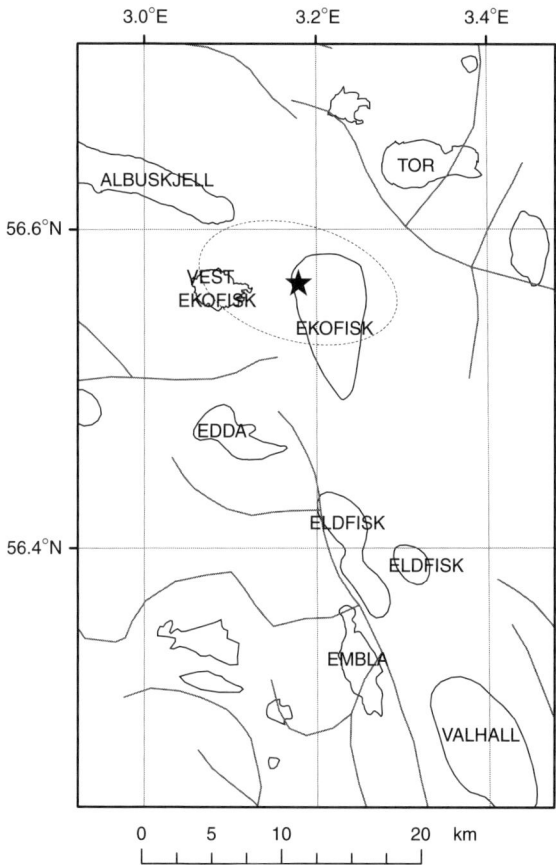

FIG. 2. Extent of the Ekofisk and neighboring fields and fault lines in comparison to the epicenter of the 7 May 2001 event indicated by a star. The dotted line represents the 90% confidence error ellipse.
Source: Figure reproduced from Ottemöller et al. (2005, Fig. 2, p. 2).

3. *Location:* Induced seismicity often occurs either directly below or directly above the reservoir (e.g. Lacq (Grasso and Wittlinger, 1990) and War-Wink (Doser et al., 1991)). This is illustrated for the Lacq field through the depth sections in Figs 3 and 4.
4. *Spatial clustering:* Nearly all studies of induced seismicity in hydrocarbon fields (e.g. Rangely (Raleigh et al., 1976), Gobles (Mereu et al., 1986), Sleepy Hollow (Evans and Steeples, 1987), Romashkino (Turuntaev and Razumnaya, 2002), and Lacq (Bardainne et al., 2006)) have confirmed that earthquakes have a pronounced tendency to form clusters or swarms. Often it is possible to associate a substantial part of the observed seismicity with a pre-existing fault or heterogeneity (e.g. Valhall (Arrowsmith and Eisner, 2006), Fashing, Imogene (Pennington et al., 1986), Lacq (Grasso, 1992b), and Shuaiba (Sze, 2005)). Figure 5 shows the

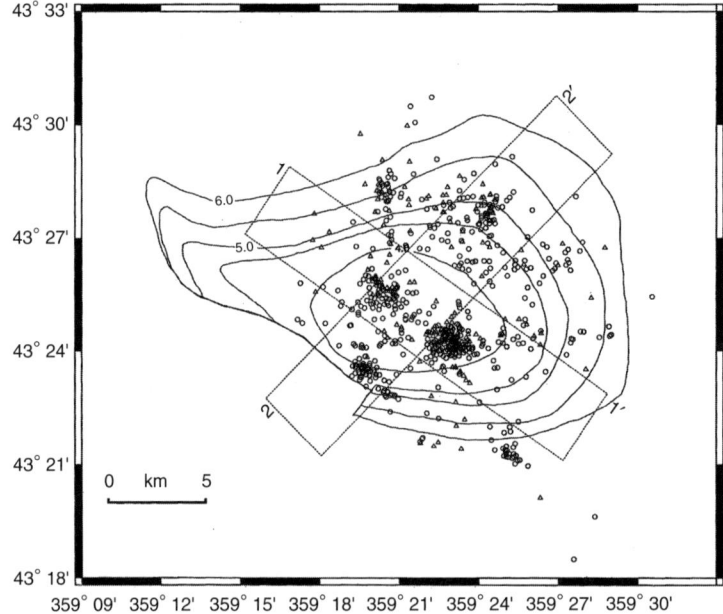

FIG. 3. Schematic representation of Lacq field (contour lines mark the depth to the top of the gas reservoir) and the induced seismicity associated with it. In accordance with Fig. 4, triangles represent events that occurred between 1976 and 1979 and circles those that occurred between 1982 and 1992. The location of epicenters was computed based on the velocity model by Guyoton et al. (1992). The boxes correspond to the location and width of the two cross sections shown in Fig. 4.
Source: Figure reproduced from Segall et al. (1994, Fig. 3, p. 15,426).

correlation between microseismicity recorded at Shuaiba field and the fault lines cutting through it.

5. *Temporal patterns:* A characteristic time lapse of several years between the beginning of production and notable increases in the seismicity level has been observed in numerous fields (e.g. Sleepy Hollow (Rothe and Lui, 1983), Cogdell Canyon Reef (Davis and Pennington, 1989), Strachan (Baranova et al., 1999), Lacq (Grasso and Wittlinger, 1990), and Wilmington (Kovach, 1974)). It should be noted, however, that quality of recording contributes to the perceived time lapse: Instrumentation is typically increased as a response to seismicity being felt by the population or observed by other networks. Thus, a potential buildup of seismicity might occur but go by unnoticed. On the other hand, in the case of extraction induced seismicity, a time-lapse between production and seismicity is in accordance with theoretical expectations from poroelastic modeling (Segall, 1985).

6. *Faulting:* The faulting in hydrocarbon fields which exhibit induced seismicity is dominated by the pre-existing stress field, since production-related stress perturbations are small in comparison. Nonetheless, production can lead both to the activation of pre-existing faults (e.g. Valhall (Zoback and Zinke, 2002), Fashing, Imogene (Pennington et al., 1986), and Rotliegendes (Van Eijs et al., 2006)) or

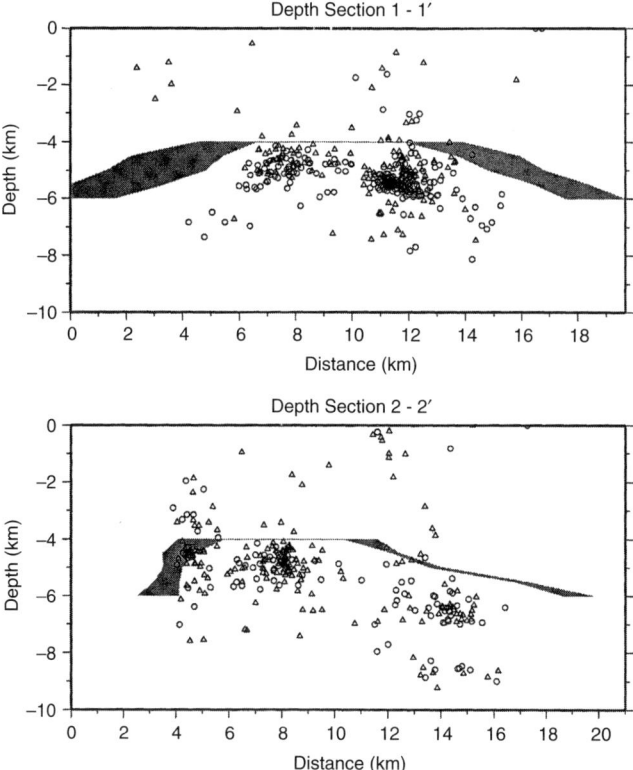

FIG. 4. Vertical cross sections through the reservoir at Lacq field. Location of the width of the cross sections is shown in Fig. 3. The gas reservoir is indicated by the regions shaded in gray. In accordance with Fig. 3, seismic activity between 1976 and 1979 is marked by triangles and that between 1982 and 1992 by circles.
Source: Figure reproduced from Segall et al. (1994, Fig. 4, p. 15,426).

to the formation of new faults as has been documented amongst others for Goose Creek (Yerkes and Castle, 1976) and Wilmington (Kovach, 1974). In the latter two cases, the formation of new faults is correlated with substantial subsidence due to extensive fluid extraction.

7. *Source Mechanisms:* The source mechanism of induced seismicity remains a topic of active research. Although many events tend to be well described by a double-couple mechanism, notable exceptions exist (e.g. Console and Rosini (1998); Talebi and Boone (1998); Foulger et al. (2004); Vavrycuk et al. (2008)). In fact, a special session of the 2008 meeting of the European Seismological Commission was devoted to non-double-couple cases.

Finally, we note that anisotropy studies have recently become a common tool in the industry to characterize dynamic changes in hydrocarbon reservoirs as reviewed by Helbig and Thomsen (2005); Crampin and Peacock (2005). In the context of fields with induced seismicity, a striking correlation between shear-wave anisotropy and subsidence

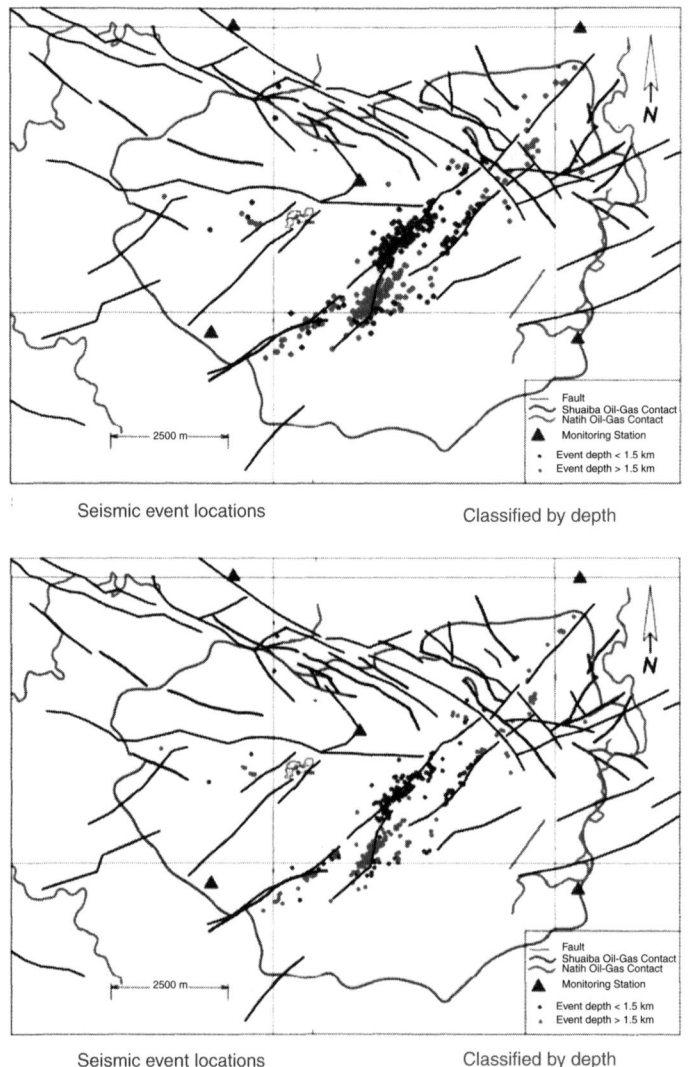

FIG. 5. Illustration of the collapsing technique for improving the localization of induced microseismicity. Shown are the locations of epicenters recorded at Shuaiba field prior to (top) and after collapsing (bottom), as well as the relevant tectonic features. Collapsing clearly reduces the scatter in localizations.
Source: Figure reproduced from Sze (2005, Fig. 2.21, p. 65).)

has been observed, in particular for Ekofisk (Van Dok *et al.*, 2003) and Valhall (Olofsson *et al.*, 2003). A plot of the azimuthal anisotropy and the measured seabed subsidence for Valhall is reproduced in Fig. 6. As pointed out by Olofsson *et al.* (2003), the azimuthal anisotropy is largest where the strain associated with subsidence is largest.

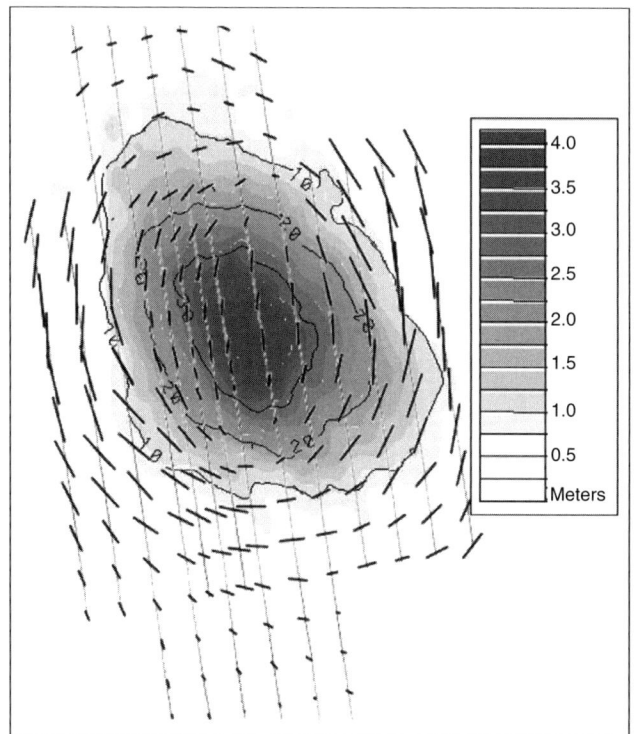

FIG. 6. Plot of the azimuthal anisotropy and the measured seabed subsidence in shades of gray (blue in the web version) for the Valhall field. Short lines indicate the S1 axes scaled by the S2 time lag.
Source: Figure as given by Olofsson et al. (2003, Fig. 1, p. 1228).

This correlation could be caused by small cracks formed during reservoir subsidence which lead to a directionally dependent shear-wave speed. We note that not only shear waves but also compressional waves have been used successfully to characterize spatial variations in hydrocarbon fields (e.g. at Valhall (Hall and Kendall, 2003)).

2.3. Regional Occurrence

According to Perrodon (1983) there are about 600 sedimentary basins worldwide out of which 400 have been drilled and 160 are (or were) used for commercial hydrocarbon production. Only 25 of these represent cumulative discoveries of substantial size (more than 1.4 billion tons) and thus concentrate about 85% of the total worldwide discoveries to date. Out of these 25, frequent incidents of induced seismicity have been clearly established only for two, namely the Permian basin, Texas, and Rotliegendes, Netherlands. However, even in these two regions not all hydrocarbon fields are similarly affected (Doser *et al.*, 1992; Van Eijs *et al.*, 2006). This observation raises the question why seismic activity is induced in certain regions and within these regions why only in certain fields and not in others.

Although numerous oil and gas fields are located in the Permian Basin, only a limited number of these fields exhibit induced seismicity in a magnitude range that is sufficient for surface recording. Clearly, this does not imply that microseismicity might not exist at depth. The relationship between induced seismicity and oil production in the Permian Basin was investigated by Doser *et al.* (1992). Based on a careful relocation of the seismic events recorded between 1975 and 1979, the authors found that seismicity appears to be correlated to a variety of effects including fluid migration, natural occurrence of overpressured fluids, tectonic activity, reservoir production, and enhanced recovery operations. For example, Doser *et al.* (1992) observed that seismicity in both the War-Wink and the Apollo-Hendrick fields tends to cluster in regions where the horizontal pressure gradient exceeds 15 MPa/km. Interestingly, seismic activity is not confined to the extent of the hydrocarbon fields: more than half the events occurring outside the Permian Basin network (or after it had ceased operation) were located between fields, which could indicate the importance of fluid movement or pressure gradients between fields (Doser *et al.*, 1992). We note that an event with magnitude $m_b = 4.7$ to $m_{bLg} = 5.0$ occurred in the region on January 2, 1992 highlighting the importance of assessing the possibility that moderate to large events could be induced by hydrocarbon production.

In the large on-land gas reservoirs in the Netherlands induced seismicity is such a commonly observed phenomenon that Dutch law requires a seismic hazard and risk estimate study for each concession since 2003 (Van Eck *et al.*, 2006). Out of the 124 producing fields, 16 have shown induced seismicity (Van Eijs *et al.*, 2006). For an overview of the hydrocarbon fields exhibiting induced seismicity in the Netherlands see Fig. 7. While natural tectonic activity occurs in the southern part of the Netherlands, induced seismicity is mostly confined to the north. An overview map of natural and induced seismicity in the Netherlands is reproduced from Van Eck *et al.* (2006) in Fig. 8. Though most of this activity is small in magnitude ($M_L \leq 3.5$), it is shallow in depth and thus occasionally causes non-structural damage (Van Eck *et al.*, 2006).

The three most active fields in terms of induced seismicity in the Netherlands are Groningen, Roswinkel, and Bergermeer fields. Groningen is the largest gas field in the Netherlands. Out of the roughly 340 induced events recorded until 2006, about 180 epicenters occurred in Groningen Field alone (Van Eck *et al.*, 2006). Most of these earthquakes seemed to occur along NW-SE trending faults at reservoir level in the northwestern part of the field. Compared to Groningen, Roswinkel, and Bergermeer are both relatively small gas fields. Interestingly, the observed seismicity patterns in these two fields are quite different from each other: while the Bergermeer field has so far only generated 4 moderately sized events ($3.0 \leq M_L \leq 3.5$), more than 36 events ranging from small ($M_L \geq 1.0$) to moderate size ($M_L \leq 3.4$) have been recorded at Roswinkel (Van Eck *et al.*, 2006).

This observation has been interpreted as evidence for the hypothesis that induced seismicity is a chaotic phenomenon (Van Eck *et al.*, 2006). A similar hypothesis was also investigated by Grasso and Sornette (1998) and led the authors to conclude that the potential seismic hazard associated with areas of large pre-existing stresses might extend well beyond the zone where natural activity is common. Van Eijs *et al.* (2006) tested the correlation between reservoir parameters, production characteristics, and the occurrence of induced seismicity in the northern part of the Netherlands in an attempt to identify which hydrocarbon field might be prone to induced seismicity before production

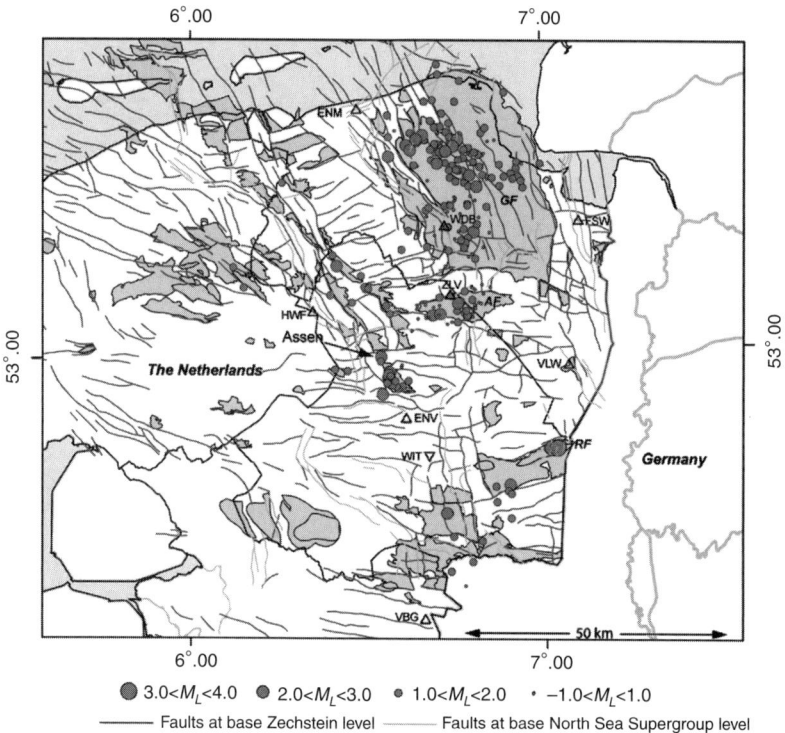

FIG. 7. Regional occurrence of induced seismicity (dark gray circles) in relation to hydrocarbon gas exploitation (gray shaded areas) and major tectonic features. The seismic station WIT (inverted triangle) and several borehole seismometers (triangles) are also indicated. Gas fields with notable induced seismicity are labeled individually: RF—Roswinkel field, GF—Groningen field, EF—Eleveld field, and AF—Annerveen field.
Source: Figure after Dost and Haak, 2008, reproduced from Van Eck et al. (2006, Fig. 4, p. 109).

begins. They concluded that three key parameters (pressure drop, fault density in the reservoir, and stiffness ratio between the seal and the reservoir rock) showed a correlation with seismicity and identified critical values that each of these parameters has to exceed in order for induced seismicity to occur. Their analysis is supported by numerical models (Mulders, 2003) and based on the observation that induced seismicity in the Netherlands is probably primarily the result of reactivation of normal faults in the reservoir. Future investigations will show whether these findings can be successfully applied to anticipating seismic hazard.

3. SEISMIC MONITORING

The monitoring of induced seismic activity is pivotal for improving our understanding of the phenomenon. Many of the advances in the continuous monitoring of microseismic

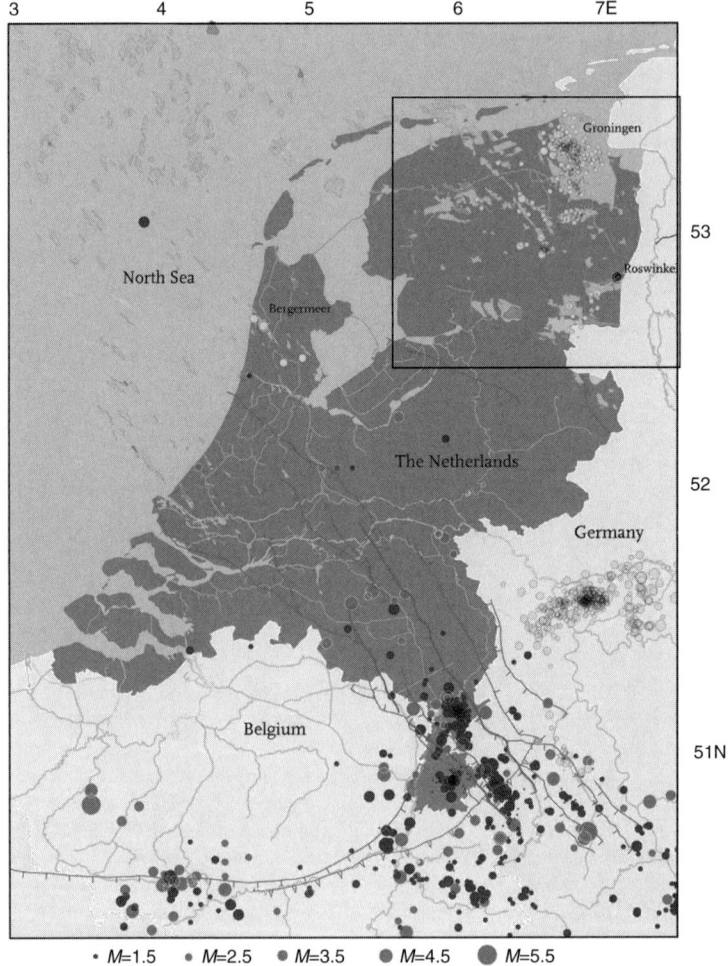

FIG. 8. Overview of the seismicity in the Netherlands and surroundings since 1900, reproduced from Van Eck *et al.* (2006). The radius of the circles is a measure of the magnitude of the event. The shading of the circles indicates the nature of the event: dark gray for natural tectonic events and light gray for earthquakes that were induced by human activity. Note that this map includes induced seismicity from both mining and hydrocarbon production. Gas fields are highlighted in gray. Mapped fault lines are plotted in gray.
Source: Figure as given by Van Eck et al. (2006, Fig. 1, p. 106).

and acoustic activity (Albright *et al.*, 1994) were pioneered in the geothermal (the earliest studies include Pearson (1981) and Albright and Pearson (1982), and many followed thereafter), and mining industry (see Gibowicz and Lasocki (2001) for a recent review of mining induced seismicity). Amongst others, it was demonstrated that the quality of the collected data was sufficient to allow tomographic imaging of the fractured volume (Block *et al.*, 1994). However, the majority of geothermal experiments have taken place in hard-rock environments, which are characterized by an efficient propagation of elastic waves. Contrary to this, most hydrocarbon reservoirs are located in sedimentary environments with less favorable properties. This led to the notion that microseismic monitoring is only applicable to reservoirs with high rock velocities (Jupe *et al.*, 1998). Nonetheless, recent studies (Jupe *et al.*, 1998; Phillips *et al.*, 1998) in environments ranging from shallow unconsolidated sands to chalks showed that this concern is not necessarily justified. Thus, interest in using induced seismicity to monitor reservoir dynamics in hydrocarbon fields has grown considerably during recent years (Jupe *et al.*, 1998; Fehler *et al.*, 2001; Maxwell and Urbancic, 2005).

Apart from hydraulic fracturing not discussed in this review, the most relevant applications of passive monitoring in hydrocarbon fields are as follows.

1. *Improving well and casing design:* Hydrocarbon production from the soft chalk formation at Valhall field led to compaction, which resulted in casing deformations in the overburden. Since production began in 1982, 28 out of 102 production wells have suffered severe tubular deformations (Kristiansen *et al.*, 2000). The observation that the locations of microseismic events correlated well with the distribution of casing deformations demonstrated the potential of continuous microseismic monitoring for improving well constructions and led to the installation of a permanent array on the seafloor with 120 km of ocean-bottom-seismometer cable and more than 10,000 geophones and hydrophones covering the field. The main purpose of the array is 4D monitoring (see Section 7), but its potential for passive monitoring is being actively investigated as well. This innovation was honored with the 12th Offshore Northern Seas Innovation Award that was awarded to BP Norway.
2. *Fault mapping:* During a temporary microseismic monitoring project of 18 days at Ekofisk field, it was possible to identify the fault pattern under the gas cloud on top of the reservoir, which obscures the structure in classical active source studies (Maxwell and Urbancic, 2005), see Fig. 9.
3. *Mapping fluid movements:* Based on a comparison of the location of the seismically active faults at Seventy-six oil field, Kentucky, in combination with additional data regarding the production history, well logs and drill tests, Rutledge *et al.* (1998) concluded that the seismically active fractures in the field had been partially drained by previous production and subsequently re-saturated with brine (water).

Two key issues in monitoring microseismicity successfully are (1) accuracy of event location and (2) reduction of ambient noise. While the latter is mostly dependent on a careful installation and sensitive, low-noise instrumentation (Maxwell and Urbancic, 2005; Kristiansen *et al.*, 2000), event localization remains challenging. Conventional earthquake location techniques are usually insufficient to get detailed information about reservoir structure, fracture orientation, and hydraulic behavior (Fehler *et al.*, 2001). Apart from the generally low magnitude of events, the limited availability of wells or boreholes requires severe compromises regarding the ideal spatial distribution of sensors.

Adequate localization of events is particularly critical to the study of induced seismicity, because the spatial correlation between earthquake occurrence and

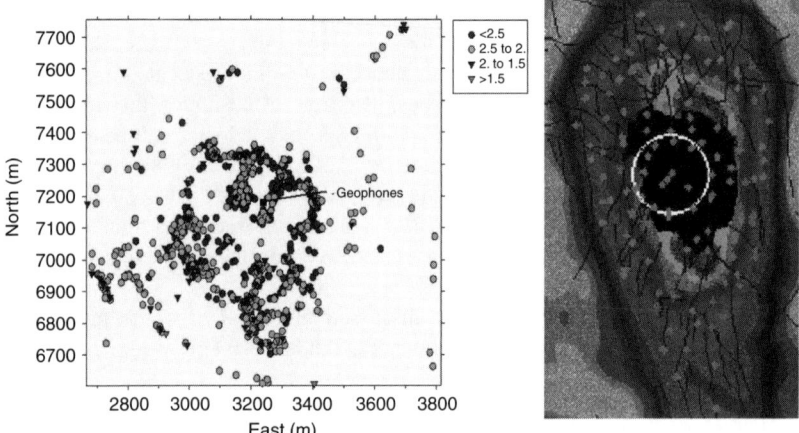

FIG. 9. Left: Microseismicity by magnitude recorded over 18 days at Ekofisk within the area indicated as a white circle on the right hand figure. Right: Structural map of Ekofisk field with the prominent gas cloud indicated in black.
Source: Figure as given by Maxwell and Urbancic (2005, Fig. 5, p. 72).

hydrocarbon production wells is usually the main indication for a possible connection between the two phenomena. There were a number of cases in which the possibility that the earthquake might have been induced was only realized after a more precise relocalization of the event. One example is the event on 20 October 2004, ($M_L = 4.5$) at Rotenburg/Soltau, Germany (Dahm *et al.*, 2007).

Since there typically is limited flexibility regarding the network geometry, recent research efforts concentrated mostly on the development of sophisticated processing techniques. Important progress in this context was made through the international research collaboration project named "More Than Cloud (MTC)" initiated in 1992 by Professor Hiroaki Niitsuma, Tohoku University. The goal of the MTC project was to develop new technology that would allow resolving more details in a "cloud" of microseismicity (Fehler *et al.*, 2001). One of the main accomplishments of the MTC project was the development of a new location technique called "collapsing" (Jones and Stewart, 1997). The key idea behind collapsing is to use the location uncertainties as a guide in processing the data.

An example of how collapsing can improve the localization of induced seismicity is the Shuaiba oil reservoir. Sze (2005) compared the results of different location algorithms to assess the stability of the epicentral relocations and to estimate reasonable bounds on the relocation errors. The microseismic data utilized by Sze (2005) in his study were collected by the Petroleum Development Oman (PDO) using a downhole geophone array between October 1999 and June 2001. The network recorded 802 events during this period. The effect of collapsing on the observed seismicity pattern as previously determined by the nonlinear grid-search technique is illustrated in Fig. 5. The top figure shows the epicenters of microseismicity before collapsing, and the bottom figure, after. The scattering of events is notably reduced. The relocated seismicity now delineates the complex zone of faults of Shuaiba field more clearly.

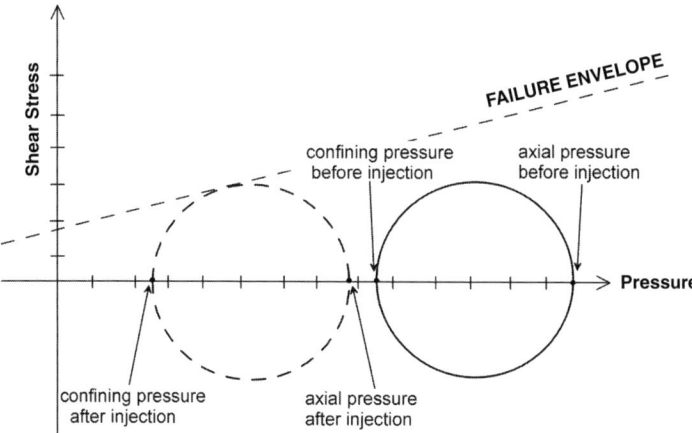

FIG. 10. Mohr diagram illustrating the effect of how injection of fluids brings a rock closer to failure.
Source: Figure as given by Sminchak and Gupta (2003, Fig. 2, p. 83).

4. Seismicity Induced by Fluid Injection

It is well known and not unexpected that fluid injection can induce seismic activity, because it decreases the effective stress: the injected fluid enters a pre-existing fracture and supports a part of the normal stress equivalent to the pressure of the fluid. As the fluid has no shear strength, the effective normal stress and the frictional resistance to sliding are lowered. If the fracture is subject to shear stress greater than the product of this effective normal stress and the coefficient of friction, the rock will slip and generate an earthquake. The argument is visualized in Fig. 10 through a Mohr circle with a Coulomb type failure criterion. In the earlier literature, the same argument is also referred to as the Hubbert-Rubey principle of effective stress.

Nicholson and Wesson (1992) listed more than 30 documented cases of potential injection induced seismicity in the USA and Canada alone, most of which are not related to hydrocarbon production but to fluid injection in different contexts. In hydrocarbon fields, massive injection of fluids is used for a variety of purposes including: (1) to replace extracted fluids in a field that has been under production for a while (secondary recovery), (2) to stimulate production by creating fractures and thereby producing new flow conduits (hydraulic fracturing), (3) to sweep fluids to producing wells, and (4) for pressure maintenance. Often the transition between these strategies is gradual.

4.1. Controlled Experiments of Fluid Injection in Hydrocarbon Fields

The Mohr-Coulomb argument of reduction of effective stress invites the conjecture that it might be possible to control earthquakes through variation of fluid injection rates. This hypothesis was tested in an experiment at Rangely oil field (Raleigh *et al.*, 1976). At Rangely field, secondary recovery of oil through injection of water at high pressures had started in 1957. The exact date of onset of seismicity is not known. Between November

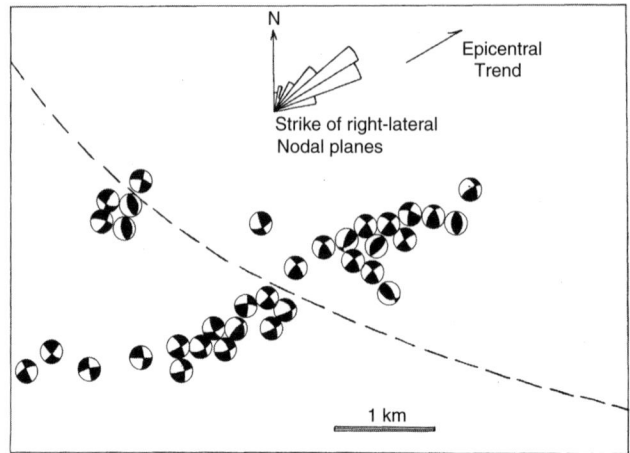

FIG. 11. Fault plane solutions for the Rangely earthquakes. Black corresponds to compressional first motion, and white to dilatational first motion. The inlet shows the azimuths of the right-lateral nodal planes. The dashed line indicates the southern boundary of the oil field.
Source: Figure reproduced from Raleigh et al. (1972, Fig. 3, p. 1233).

1962 and January 1970 the Uinta Basin Observatory recorded 976 earthquakes in its immediate vicinity (Gibbs *et al.*, 1973). In fact, Gibbs *et al.* (1973) had already suggested a connection between the fluid injections and the recorded seismicity, but did not observe a correlation between pressure and seismic activity similar to that later found by Raleigh *et al.* (1976).

The experiment at Rangely oil field began in 1969, when a seismic network of 14 short-period, vertical seismometers was installed. For one year, the network recorded the seismic activity under unaltered fluid pressure conditions. During that time, Raleigh *et al.* (1972) used hydraulic fracturing of rock in boreholes to quantify the in-situ state of stress. Given these results and their analysis of the orientation of the fault and slip direction determined from focal plane solutions of nearby earthquakes, the authors published a prediction about the critical pressure (Raleigh *et al.*, 1972). Between October 1969 and May 1973, two full cycles of increased fluid injection and backflowing were performed at Rangely. Raleigh *et al.* (1976) found that seismic activity responded promptly to changes in fluid pressure. They concluded that the experiment at Rangely had both confirmed the hypothesis that (1) earthquakes could be induced by increases in pore pressure and (2) that a simple Mohr-Coulomb argument based on the reduction of effective stress accounts for this phenomenon not only qualitatively, but also quantitatively. Figure 11 shows the fault plane solutions of the seismic activity recorded during the experiment.

Unfortunately, this type of clear correlation between induced seismicity and injection patterns is only found in the context of hydraulic fracturing experiments. The described experiment at Rangely field falls into this category (despite evidence for previous activity) as well as 76 field, Kentucky and Carthage Cotton Valley gas field, Texas, for which similar correlations have been reported (Rutledge *et al.*, 1998; Rutledge and Phillips, 2003).

4.2. Moderate Seismicity Related to Fluid Injection

Although fluid injection is typically associated with microseismicity, several events with moderate to large magnitudes have also been related to fluid injections and it is important to understand better the conditions under which moderate to large events might occur. No obvious correlation between seismicity and injection rate could be established for any of the cases exhibiting moderate seismicity. Partly, this might be due to inaccuracies in the location of hypocenters (e.g. Sleepy Hollow field (Evans and Steeples, 1987)) and missing data regarding the precise injection pattern (e.g. Gobles field (Mereu *et al.*, 1986)). Also, it is important to note that injection and depletion generally happen at different wells, which leads to a complex underground flow pattern. Thus, it is not always clear as to how the volume of fluid injected is related to the spatial variations in net pore pressure.

An additional complexity arises from the commonly observed time lapse between the beginning of fluid injections as a means of secondary recovery and the onset of seismic activity. For example, at Sleepy Hollow field, Nebraska, fluid injection for the purpose of enhancing oil production was initiated in 1966, six years after the discovery of the field (Rothe and Lui, 1983). The first indication of augmented seismic activity in the immediate vicinity of the oil field dates back to 1977. During 1979-1980, a first array of four portable seismographs was operated and in 1982 the U.S. Geological Survey installed an eight-station telemetered analog network. Rothe and Lui (1983) suggested that there probably was a connection between the earthquakes and hydrocarbon production, but were unable to rule out a tectonic cause. Evans and Steeples (1987) re-assessed their results a few years later. They noted no significant correlation between average injection pressure and earthquake occurrence, but did not consider this finding surprising. They argued that after almost 19 years of continuous injection it would be unlikely that changes in the injection pattern – without a substantial increase in overall pressure – would change the seismicity pattern.

One of the largest events in likely association with an injection operation had magnitude 4.6-4.7 and occurred at Cogdell Canyon Reef field in June 1978 (Harding, 1981; Nicholson and Wesson, 1990). Water injection as a means of secondary recovery was used at Cogdell Canyon Reef since April 1956 (Davis and Pennington, 1989). Following this event, the U.S. Geological Survey operated a local network from February 1979 to August 1981, which located a total of 20 epicenters in the Cogdell oil field. Additionally, the University of Texas/NASA seismic array recorded seismic activity in the Snyder area from April 1977 to February 1979 (Dumas, 1979). Based on this data, Davis and Pennington (1989) tested both the classic Mohr-Coulomb failure model and whether stress loading might have resulted from the weight of injected fluids. They concluded that both mechanisms are relevant, since if fluid pressures alone controlled the seismicity, it would be expected that seismic activity is concentrated in areas of high fluid pressure, which was not observed. We note that it is contentious whether the weight of fluid plays an important role in induced seismicity, because dimensional analysis suggests that this effect is comparatively small (Segall, 1985).

Another interesting case regarding the correlation between production rates and induced seismicity is that of the Romashkino oil field, Russia. From 1986 to 1992 a local seismic network operated by the Tatneftegeophysica seismic service recorded 391 local events with magnitudes up to 4.0 (Adushkin *et al.*, 2000). An epicenter map of the

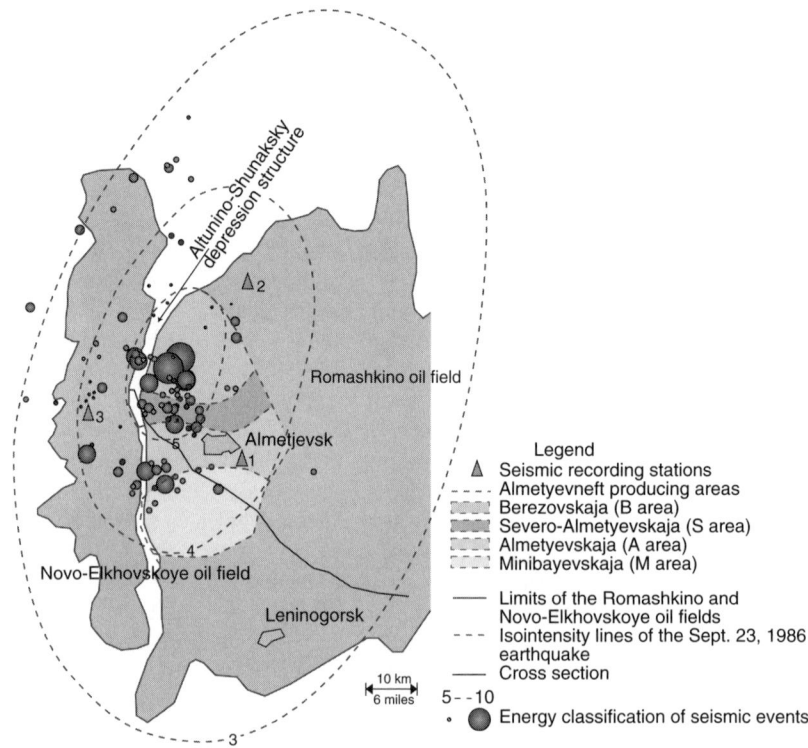

FIG. 12. Epicenter map of the seismic activity at Romashkino oil field. The radii of the circles indicate the magnitude of the event. The dashed ellipses are intensity contours for the September 1986 event. The black line shows the position of a seismic cross section and triangles represent seismic recording stations. The field is subdivided into several producing areas, which are plotted in various shades of gray.
Source: Figure reproduced from Adushkin et al. (2000, p. 8).

seismic activity at Romashkino field is shown in Fig. 12. Although a connection between production and seismic activity is not immediately obvious, Turuntaev and Razumnaya (2002) found quasi-harmonious oscillations of seismic activity, which are synchronized in time with changes of injection effectiveness. Taken over the entire observation period, seismic activity in Romashkino oil field occurred in two cycles, each lasting for about five years. Figure 13 shows a smoothed plot of the normalized seismic activity during these two cycles superimposed. Certain aspects of the two curves coincide. Most of the literature on Romashkino is in Russian; additional references can be found in Adushkin *et al.* (2000) and Turuntaev and Razumnaya (2002).

Despite these examples of injection induced seismicity, it is important to note that most hydrocarbon fields subject to similar secondary-recovery measures do not respond with increased seismic activity. Very little is known about why moderate seismic events are induced in some cases and not in others. Several attempts have been made to

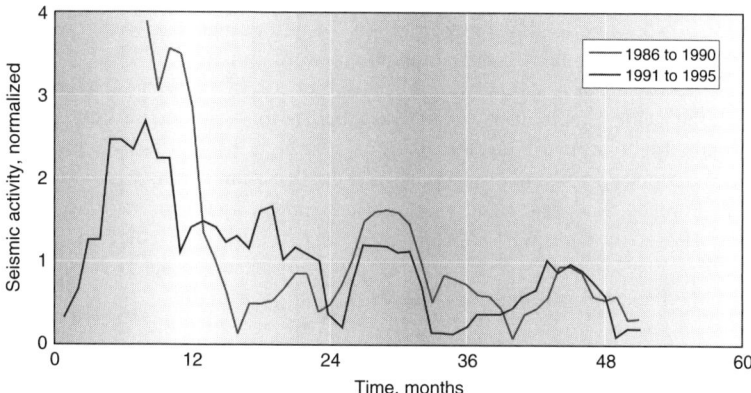

FIG. 13. Comparative superposition of the two cycles of seismic activity in the vicinity of Romashkino oil field. Each cycle lasted for about five years: the first from 1986 to 1990 and the second from 1991 to 1995. Note that both curves are normalized and have been smoothed for an easier comparison.
© 2000, Schlumberger, Ltd. Used with permission.
Source: Figure reproduced from Adushkin et al. (2000, p. 10).

generalize known case studies (e.g. Kisslinger (1976) and Davis and Frohlich (1993)), but with moderate success.

4.3. Stress Corrosion and Geochemical Processes

There is little doubt that material properties of reservoir rocks and geochemical processes play a role in understanding induced seismicity. The potential importance of stress corrosion as a mechanism relevant to induced seismicity in fluid injection environments was first pointed out by Kisslinger (1976). Laboratory experiments have proven that the time to failure is shortened substantially by increasing the water content in quartz. Similarly, the rate of crack growth under a constant load is accelerated in quartz with higher water content (Martin, 1972; Scholz, 1972). The presumed cause for this weakening or static fatigue, is the hydration of silicon-oxygen molecules, indicating that the effect might occur generally in silicate rocks and thus be relevant in the context of induced seismicity. A connection between stress corrosion and seismicity has also been hypothesized for Fashing gas field, Texas (Davis *et al.*, 1995).

Possibly the best known case for the importance of geochemical processes for reservoir management is Ekofisk field, North Sea, Norway. Although Wilmington field, located about 30 km south of Los Angeles, is infamous for being the hydrocarbon field with the largest degree of surface subsidence reported worldwide so far (9 m in 27 years), Ekofisk field has caught up over the last few years, reaching a vertical subsidence of 8.26 m in August 2002 (Ottemöller *et al.*, 2005). Discovered in 1969, Ekofisk is one of the most important oil fields in the Central graben area. The reservoir is trapped in an elongated anticline and consists of three naturally fractured, partly vertically separated chalk reservoirs at an average depth of 3000 m subsea (Key *et al.*, 1998). Although moderate events have occurred at Ekofisk (Ottemöller *et al.*, 2005), most of the deformation at Ekofisk is considered to be aseismic.

Seabed subsidence was first noted in 1984. In an attempt to reverse the impact of subsidence, full-field water flooding operations were initiated in 1987. Nonetheless, subsidence continued at a stable rate of approximately 38 cm/yr (Zoback and Zinke, 2002). At the same time, reservoir pressure began to increase (Guilbot and Smith, 2002). Detailed laboratory studies on chalk yielded the result that as hydrocarbons were produced and replaced with water, the pressured chalk began to redissolve at microscopic inter-grain contacts and redeposit in a more compact, lower porosity configuration (Torsaeter, 1984; Hermansen et al., 2000; Cook et al., 2001; Nagel, 2001; Heggheim et al., 2005). We stress that these studies primarily demonstrate the importance of geochemical effects for subsidence, not necessarily for seismicity.

5. Seismicity Induced by Fluid Extraction

Based on the argument that injection of fluid increases effective stress, it might seem counter-intuitive that seismicity could also result from the extraction of fluids. The decrease of pore pressure should decrease effective stress and thus inhibit failure. This effect plays an important role in the vicinity of pre-existing faults which are subject to a spatially heterogeneous decrease in pore pressure (Pennington et al., 1986; Nagelhout and Roest, 1997; Mulders, 2003) or if the pore pressure decrease causes a previously aseismically slipping fault to lock up and eventually release accumulated energy in a seismic event. A similar mechanism might also account for the seismicity at Grozny field, Tchetchny (see Grasso (1992b) and references in Russian therein).

But the effect of fluid extraction is more complex than can be captured by a Mohr-circle-type argument, because poroelastic stresses have to be taken into account. Evidently, reservoir rocks are composite materials incorporating a pore volume which influences their elastic response and yield stresses. The theory of poroelasticity was pioneered by Biot (1941) and a recent overview with numerous applications is given in Wang (2000). In the context of hydrocarbon reservoirs, declining pore pressures cause the reservoir rocks to contract. Because the reservoir is elastically coupled to the surrounding rocks, this contraction stresses the neighboring crust.

The substantial vertical subsidence of Goose Creek Oil field led to the first realization of the coupling between large volumes of fluid extraction and large-scale mechanical deformation (Pratt and Johnson, 1926). This case aroused the interest of Geertsma, who developed a lot of the early analytical techniques to model the effect of large groundwater withdrawals (e.g. Geertsma (1966)). However, his studies were targeted at subsidence and did not explicitly address the question of seismicity. The first to make that connection was Segall (1985). Segall derived analytic solutions for two specific symmetries: an infinitely extended horizontal reservoir (Segall, 1985, 1989) and an axisymmetric reservoir (Segall, 1992). Both models are summarized briefly below, each followed by one successful application.

5.1. Fluid Extraction from an Infinitely Extended Horizontal Reservoir

Segall (1985) modeled the oil field as a line of oil wells extracting fluid from a permeable reservoir layer embedded in an impermeable, infinite half-space. Additionally, he assumed that the depth of the permeable layer is much larger than its thickness. This setup is summarized in Fig. 14.

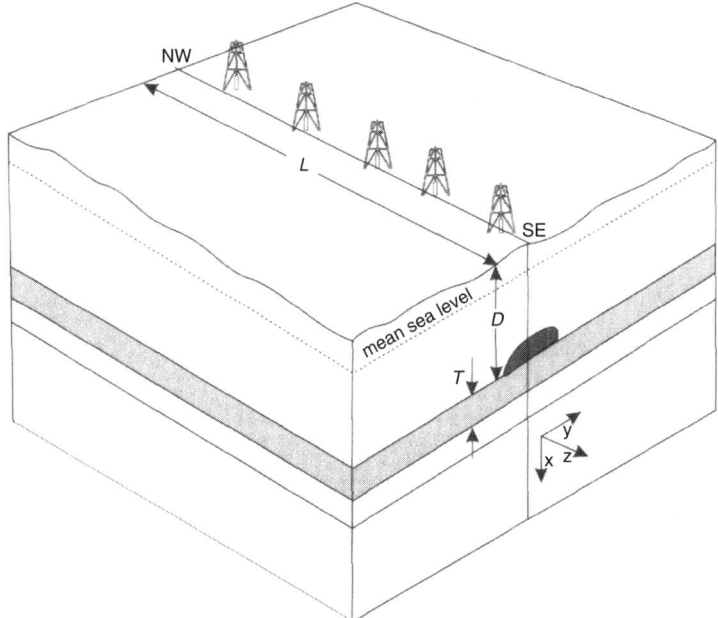

FIG. 14. Poroelastic model for fluid extraction from a horizontal layer by Segall (1985). The producing horizon of thickness T is characterized by permeability k; the half-space is impermeable ($k = 0$). Furthermore, fluid extraction is assumed to occur homogeneously along the line of wells on the surface. The lines of oil wells on the surface, the producing horizon at depth D, and the half-space are assumed to extend to infinity.
Source: Figure reproduced from Baranova et al. (1999, Fig. 8, p. 55).

Fluid extraction from this idealized geometry can only occur in the y-direction and is thus one-dimensional. As fluid is withdrawn from the reservoir, flow toward the $y = 0$ plane is induced in the producing layer. Therefore, the diffusion equation describing alterations in pore fluid mass (Rice and Cleary, 1976) reduces to

$$c\frac{\partial^2 \Delta m}{\partial y^2} = \frac{\partial \Delta m}{\partial t}, \tag{1}$$

where c is the mass diffusivity, Δm the alteration in pore fluid mass, t the time, and y the spatial coordinate. The solution to this equation subject to constant mass flux at $y = 0$ is known (Carslaw and Jaeger, 1959). From this model, three predictions follow:

1. The temporal change in pore fluid mass content Δm depends most importantly on the diffusivity within the reservoir.
2. The mean stress is compressive below the point of extraction ($y = 0$) and slightly extensional along the flanks. With increasing time since the onset of extraction the extensional zones migrate further outward and the compressive zone further down.
3. The rates of fluid extraction and surface subsidence are related linearly.

Baranova et al. (1999) applied the model by Segall (1985) to investigate whether the October 1996 event ($M = 3.9$) at Strachan gas field, Alberta, was induced by hydrocarbon production. The first indication of unusual seismic activity in north central Alberta was an earthquake of magnitude $m_b = 5.1$ which occurred on 8 March 1970. Since there are no active tectonic features in the vicinity of the epicentral area, Milne (1970) speculated about a possible connection between the earthquake and local hydrocarbon production which started in 1954. In 1980, Wetmiller (1986) deployed a seismic network consisting of one digital and six conventional seismographs and confirmed that the seismicity was induced by fluid extraction, but did not attempt to quantify the exact mechanism.

Baranova et al. (1999) investigated the event at Strachan gas field on 19 October 1996 in more detail. The authors concluded that the model is well suited to describing the phenomenon, because:

1. The measured and theoretically expected focal mechanisms of the October 1996 event corresponded well.
2. There was the expected characteristic time lag between the beginning of production in 1970 and the onset of seismicity in 1975/1976 during which enough stress perturbations built up to induce seismic events.
3. There was a correlation between the occurrence of seismic events and the production rate at the Strachan D3-A pool.

5.2. Fluid Extraction from Axisymmetric Reservoirs

Segall (1992) presented an analytic solution for the stresses and surface deformation induced by fluid extraction from disk-shaped, axisymmetric reservoirs. His study builds on previous work by Geertsma (1973). Segall et al. (1994) demonstrated that the original model can easily be extended to axisymmetric dome-shaped reservoirs.

The starting point is the four governing poroelastic equations in the pure compliance formulation. In terms of displacements u_i and pore fluid mass Δm, they take the form

$$\mu \nabla^2 u_i + \frac{\mu}{1 - 2\nu_u} \frac{\partial^2 u_j}{\partial x_i \partial x_j} - \frac{B K_u}{\rho_0} \frac{\partial \Delta m}{\partial x_i} = 0 \quad \text{for } i = 1, 2, 3. \tag{2}$$

$$c \nabla^2 (\Delta m) = \frac{\partial \Delta m}{\partial t}, \tag{3}$$

where μ denotes the shear modulus, ν_u the Poisson ratio, B Skempton's pore-pressure coefficient, K_u the bulk modulus, and ρ_0 the fluid density in the reference state. In Eq. (2) and the following expressions, the subscript u refers to the undrained state.

It is important to note that the change in fluid mass content and thus also the pore pressure p enters the equations equivalent to a body force f_i

$$\mu \nabla^2 u_i + \frac{\mu}{(1 - 2\nu)} \frac{\partial^2 u_j}{\partial x_i \partial x_j} - \alpha \frac{\partial p}{\partial x_i} + f_i = 0, \tag{4}$$

where α is the Biot pore pressure coefficient. Due to this correspondence it is possible to use the elastostatic Green's function $g_i^k(\vec{x}, \zeta)$, which is the displacement in the

i-direction at $\vec{x}$ due to a body force in the k-direction at ζ, to calculate the displacements induced by gradients in pore pressure. Assuming a localized pore pressure disturbance, Segall (1992) obtained

$$\mu u_i(\vec{x}) = \alpha \int_V p(\zeta) \frac{\partial g_i^k(\vec{x}, \zeta)}{\partial \zeta_k} dV_\zeta. \tag{5}$$

Note that the expression $\frac{\partial g_i^k(\vec{x}, \zeta)}{\partial \zeta_k}$ describes a sum of force couples acting along three orthogonal axes; in other words, a center of dilatation. The problem is thus reduced to finding the displacement Green's function for a center of dilatation defined as

$$g_i(\vec{x}, \zeta) \equiv \frac{\partial g_i^k(\vec{x}, \zeta)}{\partial \zeta_k} \tag{6}$$

(Love, 1944). The calculation of the appropriate Green's function is facilitated by the prior work of Goodier (1937), who showed that the thermoelastic equations could be reduced to the Poisson equation by introducing a displacement potential Φ such that

$$\frac{\partial \Phi}{\partial x_i} = u_i. \tag{7}$$

Mindlin and Cheng (1950) showed that Goodier's method can be extended to a homogeneous half-space. Thus, the problem is reduced to finding the solution to the Poisson equation for radial symmetries (Segall, 1992).

The resulting vertical Green's function for a ring of dilatation at radius ρ and various depths d is shown in Fig. 15. For a shallow source, a radially symmetric increase in pore pressure leads to a ring of vertical uplift at the source radius. With increasing depth of the pore-pressure source, the ring of uplift broadens and decreases in amplitude until it eventually transforms into a central bulge (Fig. 15). Beyond that depth, the deformation at the free surface caused by a toroidal source becomes indistinguishable from that of a point source (Segall, 1992). The vertical displacement shown in Fig. 16 is the convolution of the vertical Green's function with the pressure distribution.

From Green's functions, Segall (1992) calculated the perturbations of radial, vertical, hoop, and shear stresses. Knowledge of these perturbing stresses allows determining whether hydrocarbon production brings a fault of a given orientation closer to failure. However, knowledge of the tectonic background stresses is required to ultimately determine whether a given perturbation is significant enough to cause earthquakes. Thus, an identification of areas of increased seismic risk due to hydrocarbon production is strictly only possible if the natural stress field can be constrained (Segall, 1992).

Summarizing, the prediction of the effects of fluid extraction for axisymmetric reservoirs is as follows.

1. A linear relationship between decrease in pore pressure and surface subsidence.
2. The amount of subsidence depends on the elastic and poroelastic properties of the reservoir. For stiff reservoirs (e.g. Lacq), only a few centimeters of surface

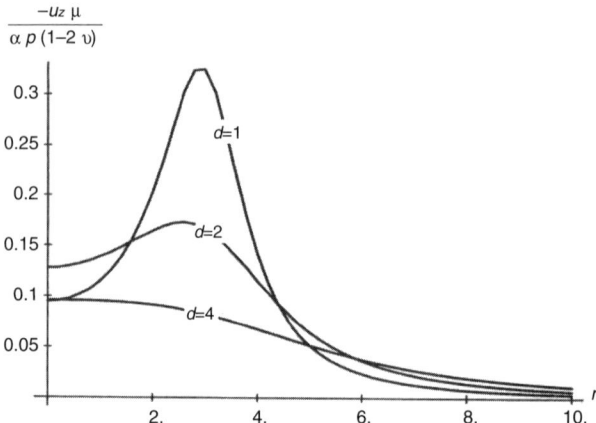

FIG. 15. The vertical Green's function at the free surface corresponding to a radially symmetric pore pressure increase of magnitude p with radius $\rho = 3.0$ and depths $d = 1, 2$, and 4. The result shown is normalized by factor $-\alpha p(1 - 2\nu)/\mu$.
Source: Figures reproduced from Segall (1992, Fig. 2, p. 544).

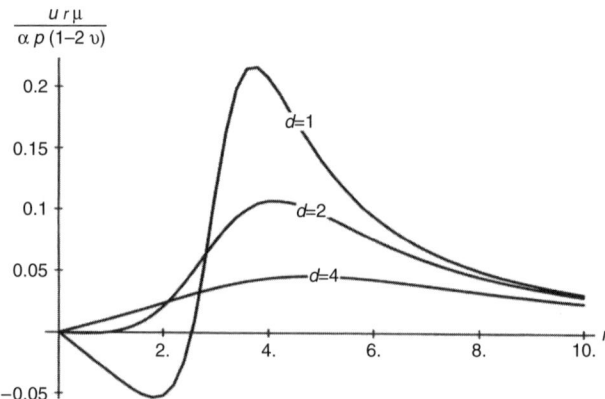

FIG. 16. The vertical surface displacement at the origin ($r = 0$) as a function of the radius ρ of the pore pressure source with magnitude p located at depths $d = 1, 2$, and 4. The normalization is the same as in Fig. 15.
Source: Figure reproduced from Segall (1992, Fig. 3, p. 545).

subsidence are expected. In weaker reservoirs where subsidence is large (e.g. Wilmington (Kosloff et al., 1980a,b), Ekofisk (Teufel et al., 1991), Costa Oriental near Lake Maracaibo, and Venezuela (Xu, 2002)), nonlinear effects become important (Kosloff et al., 1980b).

3. For a shallow reservoir, a radially symmetric decrease in pore pressure leads to a ring of vertical depression. For a reservoir of intermediate depth, the depression

is broader and decreases in amplitude toward the center of the reservoir. Finally, a deep reservoir is characterized by a central bulge of deformation. Beyond a certain depth, the deformation at the free surface caused by a toroidal source becomes indistinguishable from that of a point source. In that case, the radial displacements are oriented inward.

Segall *et al.* (1994) applied this model to the Lacq gas field, Aquitaine, France. The Lacq gas field in southwestern France is one of the best documented cases of seismicity induced by the extraction of fluids (Feignier and Grasso, 1990; Grasso and Feignier, 1990; Grasso and Wittlinger, 1990; Maury *et al.*, 1992; Guyoton *et al.*, 1992; Grasso *et al.*, 1992; Herquel and Wittlinger, 1994; Lahaie and Grasso, 1999). Since 1969, hundreds of shallow earthquakes with small to moderate magnitudes have been recorded by the French National network in its vicinity. Additionally, a local seismic network was operated and confirmed that the epicenters of the events coincided well with the extent of the gas field (Grasso and Wittlinger, 1990). Information about surface subsidence at Lacq has been obtained through repeated leveling. The first leveling was done well before the beginning of production in 1887, and then again in 1967, 1979, 1989, and 1990 (Segall *et al.*, 1994).

Despite the low permeabilities of the rocks in the producing horizon, the pore-pressure distribution within the central reservoir has been largely uniform, which Segall *et al.* (1994) interpreted as an attestation of the high fracture permeability in the reservoir. This recognition also led to the concentration of production wells at the center of the reservoir, which yields an approximately axisymmetric symmetry in agreement with the model assumptions. This makes Lacq an ideal test for the poroelastic model and a successful one: it could both explain the observed linear relationship between pore pressure in the gas reservoir and maximum subsidence (Fig. 17) and reproduce the measured surface subsidence satisfactorily (Fig. 18). Evidently, the fit of theoretical and experimental subsidence is more accurate if the domal structure of the reservoir is accounted for. The tendency of the observed subsidence to be concentrated more at the center of the field might be an indication for the presence of inelastic effects (Segall *et al.*, 1994).

Segall *et al.* (1994) calculated the perturbation in the pre-existing tectonic stress field, which results from fluid extraction from the gas reservoir and concluded that changes in Coulomb stress as small as 0.1 MPa were sufficient to induce earthquakes. Thus, the state of stress is clearly still determined by tectonic stresses and detailed knowledge of the natural stress state is essential for anticipating which parts of a hydrocarbon field might be most prone to induced seismic activity. Adopting a simple Coulomb failure criterion and assuming a thrust faulting environment with the largest stress component oriented parallel to the Pyrenees, Segall *et al.* (1994) attempted a prediction where induced seismicity should be observed predominantly. Figure 19 shows the resulting Coulomb failure stress and observed earthquakes. Theoretically, all earthquakes should cluster in the center of the dark shaded area.

Despite the progress made by Segall *et al.* (1994), the details of the spatial and temporal distribution of seismicity at Lacq remain puzzling. As pointed out by Grasso (1992b), it failed to offer an explanation for the decrease of seismic activity since 1980 or for spatial clustering (Grasso and Wittlinger, 1990). Feignier and Grasso (1990) and Grasso and Feignier (1990) looked at the role of pre-existing discontinuities of either

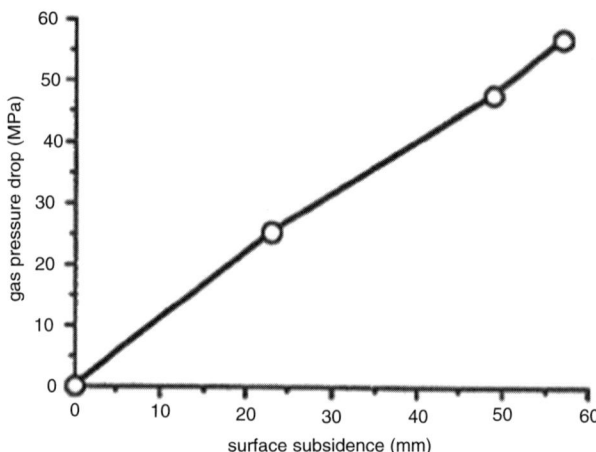

FIG. 17. Plot of the gas pressure decline in MPa over the maximum observed surface subsidence in mm at Lacq gas reservoir, Aquitaine, France. This observation confirms the model prediction of a linear relationship between subsidence and pore pressure.
Source: Figure after Segall et al. (1994, Fig. 7, p. 15,429).

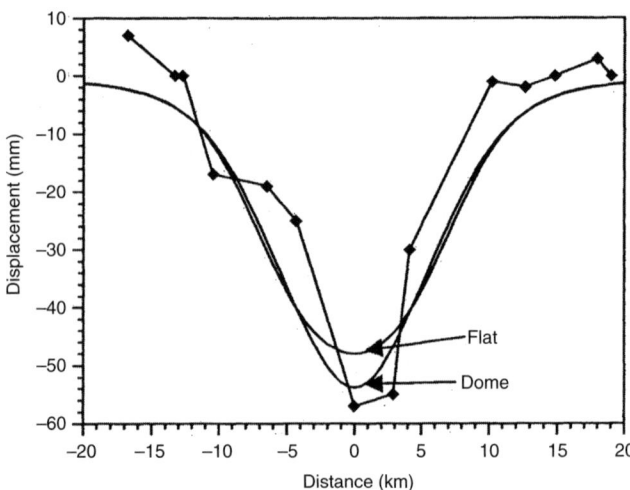

FIG. 18. Comparison of measured vertical displacement and model predictions for both a flat axisymmetric reservoir and an anticlinal dome structure at Lacq gas reservoir, Aquitaine, France for the period 1887-1989.
Source: Figure after Segall et al. (1994, Fig. 9, p. 15,432).

tectonic or lithological nature. Furthermore, a recent study by Bardainne *et al.* (2006) showed that at least part of the seismic activity at Lacq is likely due to fluid injection.

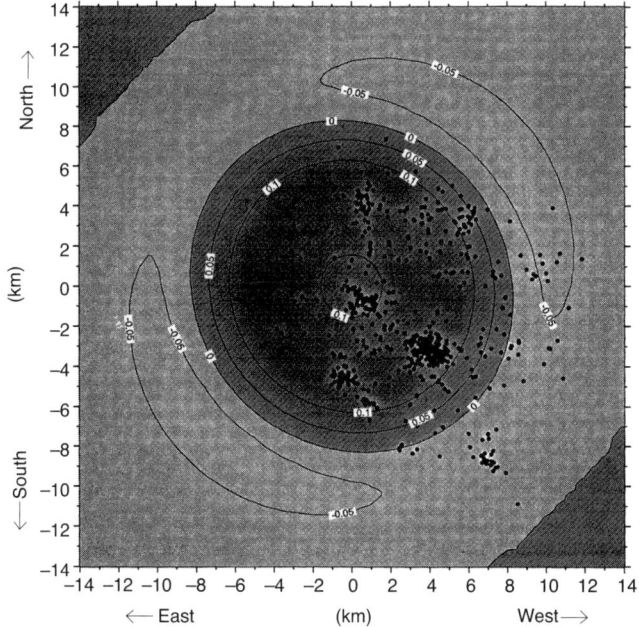

FIG. 19. Superposition of the induced and the assumed pre-existing tectonic Coulomb stress state in a horizontal section at a depth of 4.5 km. The pre-existing stress field is constructed with respect to the Pyrenees: the largest principal stress component was chosen to be perpendicular to the Pyrenees (N45°E) and the smallest is vertical. Dark shading indicates regions where reverse slip is promoted; light shading indicates areas where reverse slip is inhibited as a consequence of fluid extraction. Theoretically, earthquakes are expected to cluster in the center of the dark shaded area.
Source: Figure after Segall et al. (1994, Fig. 13, p. 15,435).

5.3. Fluid Extraction in the Vicinity of Pre-existing Faults

Poroelastic modeling usually assumes that the reservoir is embedded in a homogeneous medium. This is clearly rarely a realistic assumption. Pennington *et al.* (1986) suggested a different mechanism through which fluid extraction could lead to earthquakes on pre-existing faults in immediate contact with the reservoir. The key idea of their model is that a decrease in pore pressure strengthens a fault by increasing the effective normal stress acting on it and thus increases its strength. As a consequence, a barrier to slip develops in the immediate vicinity where the fluid is being extracted. Stress builds up along these locked portions of the fault, because strain accumulates either due to differential compaction or continued aseismic slip of nearby portions of the fault. Eventually, the accumulated stress will exceed the strength of these asperities and result in an earthquake. This process repeats itself as long as the fault is active and the pore pressure continues to decrease. An important prediction of this model is that as fluid withdrawal continues, the magnitude of future earthquakes is expected to increase (Pennington *et al.*, 1986).

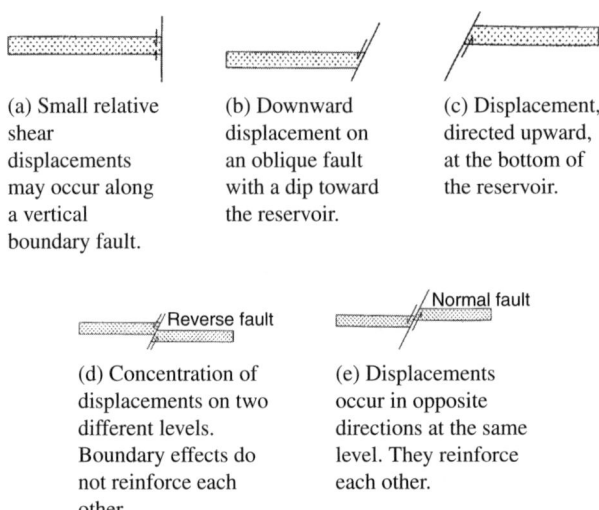

FIG. 20. Impact of geometrical effects on a fault as suggested by Roest and Mulders (2000)
Source: Figure reproduced from Roest and Mulders (2000, Fig. 2, p. 334).

Pennington et al. (1986) argued that their model might account for the seismicity observed at Imogene oil field near Pleasanton, Texas, and Fashing gas field near Fashing, Texas. Both fields are seismically active without showing massive subsidence, are characterized by structural traps formed by the offset of porous limestone along normal faults, and have experienced substantial pressure drops since the beginning of production: At the initiation of production at the Imogene oil field in 1944, fluid pressure within the field was at 24.6 MPa. By 1973 it had dropped to 14.6 MPa. At the Fashing gas field, production began in 1958 and at that time, the fluid pressure within the reservoir was approximately 35.2 MPa. In 1983, it had decreased to about 7.1 MPa near the fault (Pennington et al., 1986).

The largest events in the prior seismically quiescent area of South Texas occurred at Fashing in July 1983 ($m_{bLg} = 3.4$) and at Pleasanton in March 1984 ($m_{bLg} = 3.9$). On April 9th, 1993, another large earthquake ($m_{bLg} = 4.3$) shook south-central Texas. Its vicinity to the Fashing gas field generated new interest in the question of induced seismicity in the field. A study by Davis et al. (1995) confirmed the conclusion by Pennington et al. (1986) that seismicity in both the Fashing and the Imogene field is induced by fluid extraction. Additionally, Davis et al. (1995) pointed out that production at Fashing and Imogene might be affected by the geochemistry in the fields. In fact, at Fashing, water has been reported to leak through the trapping fault and encroach into the gas field.

The ideas by Pennington et al. (1986) have been applied and carried further, both generally to include geometrical effects (Nagelhout and Roest, 1997; Roest and Kuilman, 1994) and specifically to model induced seismicity in the Netherlands (Mulders, 2003). Figure 20 gives an overview of the geometrical effects in the vicinity of a pre-existing fault (Roest and Mulders, 2000). In order to study the influence of various parameters on stress development and fault slip, Mulders (2003) computed 2D and 3D geomechanical models of gas reservoirs based on the finite element software package DIANA. The

FIG. 21. A vertical section through an analog model by Odonne *et al.* (1999). The depletion of the reservoir is represented by the deflation of a balloon. Steep reverse faults form a cone shape during depletion. Surface subsidence is concentrated at the center of the field.
Source: Figure as given by Odonne et al. (1999, Fig. 4, p. 113).

geometry of the reservoir is derived from field data and is representative of Groningen, Annerveen, and Bergermeer fields, Rotliegendes, Netherlands.

5.4. Analog Models of Faulting Induced by Fluid Extraction

Odonne *et al.* (1999) designed an analog model to get experimental insights into the geometry and location of brittle deformation associated with depleting hydrocarbon reservoirs. The authors represented the depleted reservoir either by a latex balloon or by undercompacted ground sand, which is embedded in a sand-silicone box. Vertical sections cut into the model after the cessation of subsidence revealed three distinct areas: a contracted center, an extended ring, and a fixed periphery. Furthermore, Odonne *et al.* (1999) observed reverse faults with high dips, which is in accordance with the observations at Strachan (Wetmiller, 1986) and Lacq field (Feignier and Grasso, 1990). In the analog model, these steep reverse faults were located at the boundary of the reservoir and delimited a cone-shaped volume with an upward-directed apex as shown in Fig. 21. The curved profile of the fault seemed to depend on the ability of the fault to cut the surface of the model; in other words, in the case of the analog model on the thickness of the sand cover (Odonne *et al.*, 1999).

Based on their results, Odonne *et al.* (1999) offered a new interpretation of the analytic model by Segall (1989). The main change from Segall's model is that Odonne *et al.* (1999) suggested choosing the steep nodal plane at depth instead of the gentle one. Furthermore, they continued this steep plane upward toward the surface, resulting in a cone-shaped fault. Both the original cross section by Segall (1989) and the suggested modified cross section by Odonne *et al.* (1999) are shown in Fig. 22.

6. THE CONTROVERSY SURROUNDING MAJOR MIDCRUSTAL EARTHQUAKES

All cases of induced seismicity in hydrocarbon fields presented so far have exhibited small or moderate seismicity. A similar range of magnitude has been observed for

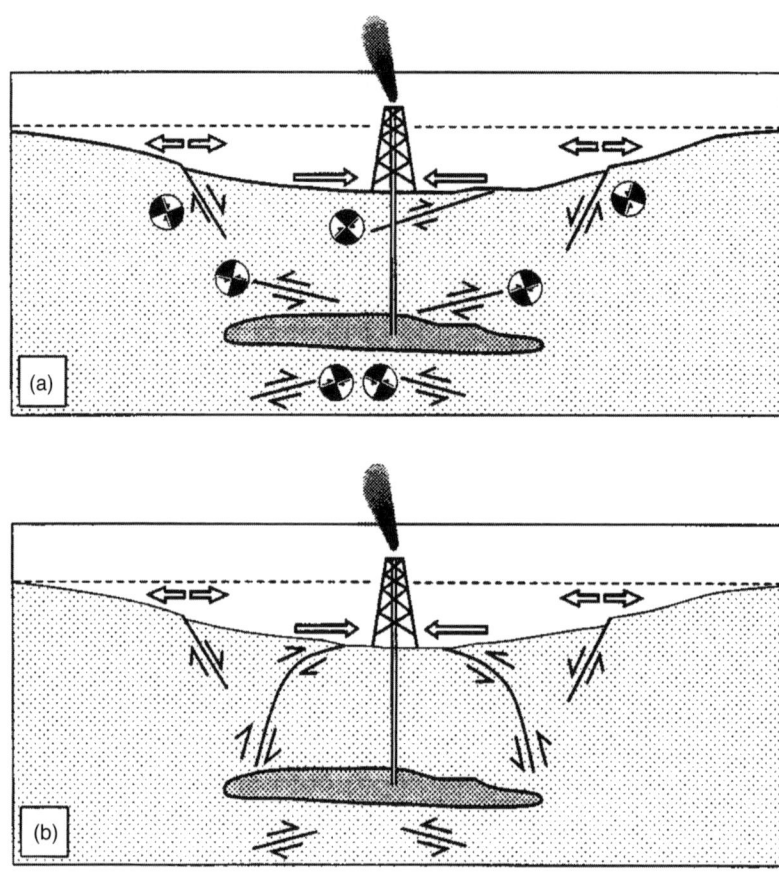

FIG. 22. Comparison of the schematic cross section summarizing surface deformation, faulting, and fault mechanisms associated with fluid extraction after Odonne *et al.* (1999). The top cross section (a) is reproduced from Segall (1989) and the bottom cross section (b) is the modification suggested by Odonne *et al.* (1999) based on their analog models. White arrows indicate horizontal displacement and black errors the sense of faulting. The focal mechanism associated with the faults are shown as beach balls.
Source: Figure as given by Odonne et al. (1999, Fig. 4, p. 113).

seismicity caused by other human activities such as mining. However, there are two examples of destructive earthquake sequences which occurred in the immediate vicinity of oil fields and which were not related to a previously known fault. These have given rise to speculations about the possibility that hydrocarbon production might cause destructive earthquakes. Apart from their size, a common feature of these earthquake sequences is that they occurred at midcrustal depths (around 10 km). Thus, the question about a possible connection to hydrocarbon production is closely related to the question of how induced stress perturbations could propagate to these depths.

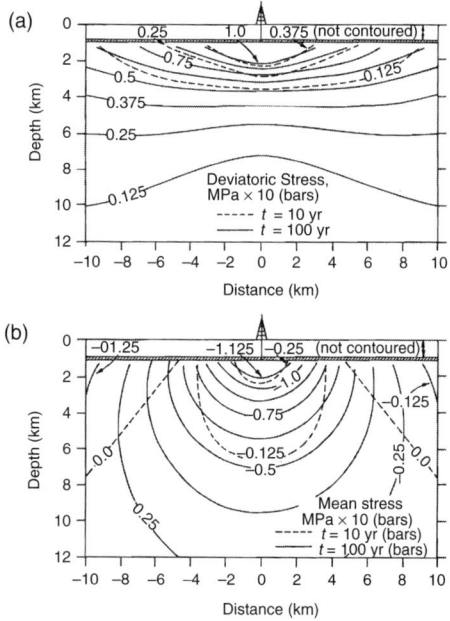

FIG. 23. Deviatoric (a) and mean (b) stress resulting from fluid extraction from a single horizontal layer of infinite extent. The contours are lines of constant stress change after 10 years (dashed) and 100 years (solid), respectively. Compressive stresses are labeled negative and extensional stresses positive.
Source: Figure reproduced from Segall (1985, Fig. 9).

6.1. Earthquakes in the Vicinity of Coalinga

On 2 May 1983, an earthquake of magnitude $M_L = 6.7$ occurred about 12 km northeast of the town of Coalinga, California. It was located 35 km northeast of the San Andreas fault, in a region that had previously been known only for scattered seismicity. The event was not associated with any known or suspected active fault. Its proximity to two major producing oil fields, Coalinga Eastside and the Nose area of the Coalinga East Extension, led to speculation of a possible connection between the event and oil field operations (Bennett and Sherburne, 1983).

Segall (1985) and Segall and Yerkes (1990) analyzed how production-related decreases in pore pressure alter the state of stress at various depths. Based on his previously developed poroelastic model for the plane strain case (Segall, 1985), he calculated that in the case of Coalinga the stress induced by fluid extraction is only 0.02 MPa at hypocentral depth. Figure 23 shows the mean and deviatoric stress due to fluid extraction from a single horizontal layer based on poroelastic modeling. As an independent validation for the model, Segall (1985) used it to model the surface subsidence at Coalinga oil field and found a very good fit of theory and observations. The theoretically predicted surface subsidence for various time intervals after onset of production is reproduced in Fig. 24. Pore pressure changes calculated on the basis of this

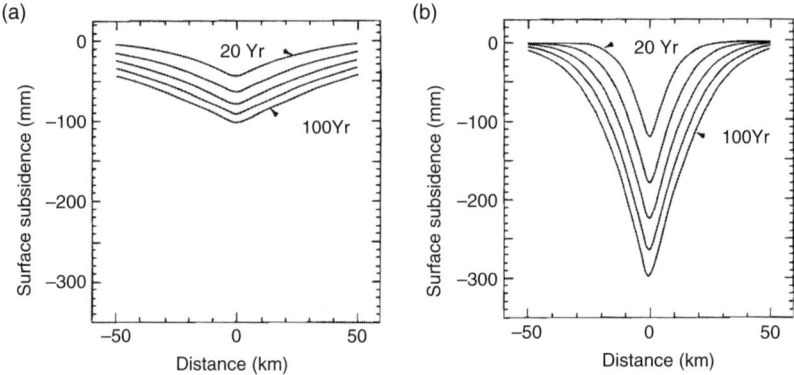

FIG. 24. Expected surface subsidence (mm) in time increments of 20 years starting 20 years after onset of production and ending 100 years later, as computed from the poroelastic model by Segall (1985). The left figure (a) is computed based on a diffusivity c, which is a tenth of the diffusivity used in the right figure (b), highlighting the great importance of that parameter. The other quantities are specified in Segall (1985).
Source: Figure reproduced from Segall (1985 Fig. 8).

model, also agreed well with observed changes (Segall and Yerkes, 1990), thus lending credibility to the poroelastic approach.

The U.S. Geological Survey initiated a detailed study of the event (Rymer and Ellsworth, 1990). This report came to the conclusion that the earthquake was closely associated with a fault zone concealed beneath folds developed along the structural boundary between the Coast (Diablo) Ranges and the San Joaquin Valley. A mechanism related to hydrocarbon production was ruled out for the following three reasons.

1. The shale units which underlie the oil-producing horizon have extremely low permeability and act as a hydraulic isolation between the oil field and the earthquake's focal region.
2. Despite the usage of fluid injection in the course of secondary recovery projects, the net effect of oil production was a 50-percent decline in reservoir fluid pressure between 1938 and 1983 (Segall and Yerkes, 1990).
3. Except for one aftershock on June 11, all other aftershocks and the main shocks had focal depths of 9 to 12 km. Any pore pressure disturbance resulting from oil field operations would have had to penetrate at least 8 km during the 45 years since operation of the field started in 1938. However, Segall and Yerkes (1990) estimated that the pressure disturbance will only be able to penetrate 100 m in 45 years. There remained, of course, the possibility of a highly fractured zone linking the oil field and the focal region. However, as argued by Yerkes *et al.* (1990), considerable evidence supports the conclusion that the rocks below the oil-bearing strata hydraulically isolate the oil field.

After the question of a possible connection between oil field operations and midcrustal earthquakes had seemingly been answered for the 1983 Coalinga earthquake, new interest arose after the occurrence of two other major events in the vicinity of Coalinga, the 1985 Kettleman North Dome and the 1987 Whittier Narrows event, that were also located directly beneath important oil fields (McGarr and Simpson, 1997).

McGarr (1991) highlighted the similarity between the three events and suggested a new mechanism that might account for a possible connection between their occurrence and oil field production. He argued that net extraction of oil and water reduces the average density of the upper crust, thus causing an isostatic imbalance. The deformation induced by this imbalance results in an increasing load on the seismogenic layer. As pointed out by McGarr himself, it is not so much a question whether the removal of mass from the crust creates an isostatic imbalance, but whether the resulting deformation would occur on a time scale corresponding to oil field operations.

The following similarities between the three events led him to believe that isostasy might account for the connection between their occurrence and oil production: (1) All three events occurred beneath major oil fields and the aftershocks also clustered in the general area of the fields (Segall, 1985; Ekström *et al.*, 1992; Davis *et al.*, 1989), (2) the focal mechanisms of all three events are predominantly of thrust type (Eaton, 1990; Ekström *et al.*, 1992; Hauksson *et al.*, 1988), (3) the hypocenters were located at midcrustal depths and separated from the producing formation, (4) they all have comparable magnitudes, (5) the ratio of net fluid production to total seismic moment is similar, and (6) the three oil fields in question, Coalinga, Kettleman, North Dome, and Montebello, are situated on anticlines that have undergone recent uplift as a consequence of horizontal tectonic compression of the crust.

Based on his mechanism, McGarr (1991) expected that other Californian oil fields might be candidates for induced midcrustal seismicity. These are oil fields that are both located along the Coast Range-Sierran Block boundary zone and characterized by high net liquid extraction. In particular, he named Midway-Sunset, Belridge South, Elk Hills, Cymric, and McKittrick oil fields (McGarr, 1991). Apart from microseismicity related to hydraulic fracturing, no large or unexpected earthquakes have been associated with these oil fields until today. Finally, we note that Segall (1985) argued that the stress change due to mass redistribution is small in comparison to poroelastic effects based on dimensional analysis.

6.2. The Gazli Earthquake Sequence

From 1976 to 1994 an unusual earthquake sequence was recorded in the Gazli gas field located in Central Asia about 100 km northwest of Bukhara, Uzbekistan. The first destructive event occurred on 8 April 1976 with magnitude $M_S = 7.0$. It was followed shortly afterward, on 17 May 1976, by a second similarly sized event ($M_S = 7.0$) approximately 20 km west of the first event. A medium sized event ($M_S = 5.7$) followed on 4 June 1978. A few years later, on 19 March 1984, a fourth major event with $M_S = 7.0$ was located 15 km southwest of the second event in 1976 (Eyidoğan *et al.*, 1985). All three earthquakes, including some of their aftershocks, were widely felt and caused major damage in the town Gazli. Their exact locations, rupture kinematics and focal mechanisms remain a subject of debate (Eyidoğan *et al.*, 1985; Amorèse *et al.*, 1995; Amorèse and Grasso, 1996; Bossu *et al.*, 1996; Bossu and Grasso, 1996). The main events of this sequence and the available focal mechanisms are shown in Fig. 25. For details on the displayed events see Table 2.

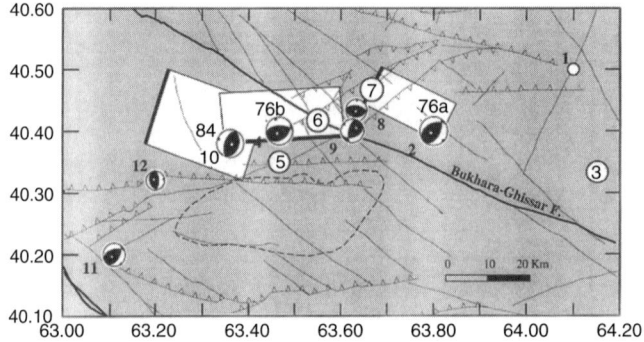

FIG. 25. Major events that occurred in the vicinity of Gazli field located by the Uzbek network between 1975 and 1994. Focal mechanisms are displayed when available with diameters representing the energy class of each event. The details of the events are listed in Table 2. The periphery of the gas field is indicated by the dotted line. Structural features were derived from LANDSAT and SPOT images (see Bossu *et al.* (1996)) for details. The white rectangles represent the geodetic fault planes of the largest three events: rectangle 76a corresponds to event 2, 76b to event 4 and 84 to event 10.
Source: Figure reproduced from Bossu et al. (1996, Fig. 3, p. 962).

TABLE 2. Parameters of the largest events in the Gazli region from Bossu *et al.* (1996, p. 962)

Event	Date	Time (UT)	Depth [km]	M (ISC)	Source
1	04/13/75	15:02	10	4.6b	
2	04/08/76	02:40	20	7.0s	Amorèse *et al.* (1995)
3	04/08/76	02:59	20	6.4s	
4	05/17/76	02:58	13	7.0s	Amorèse *et al.* (1995)
5	05/18/76	04:16	8	4.5s	
6	05/28/76	14:05	5	4.8s	
7	04/21/77	13:38	10	4.9s	
8	07/14/77	05:49	5	5.4s	CMT
9	06/04/78	19:30	15	5.7s	Uzbek network
10	03/19/84	20:28	15	7.0s	Amorèse *et al.* (1995)
11	03/20/84	06:28	15	4.2s	CMT
12	05/25/94	07:42	5	5.2b	Uzbek network

The giant Gazli gas field with an estimated size of 456 billion m^3 was discovered in 1956. Gas is pumped from a shallow reservoir at 2 km depth, although the gas must originate considerably deeper. The field is characterized by several fold and fault structures, which are part of the Bukhara-Gissar system continuing further east into a seismically more active region (Bossu, 1996). Simpson and Leith (1985) were the first to suggest a connection between the earthquake and hydrocarbon production, based on the following rationale.

1. Previously to this earthquake sequence, the region had been seismically inactive since medieval ages (Bossu, 1996).
2. The occurrence of two magnitude 7 events, followed by a third event 8 years later does not follow a typical aftershock pattern.
3. Hydrocarbon production had led to a huge decrease in pressures in the Gazli field since the beginning of its operation.
4. Source modeling of the 1984 event indicates that the rupture propagated downward (Eyidoğan *et al.*, 1985), which Simpson and Leith (1985) interpreted as evidence of an unusual near-surface stress distribution.

On the other hand, Eyidoğan *et al.* (1985) interpreted the Gazli earthquake sequence as an indication for the presence of a major fault zone to the north of Gazli. The authors argued that the observed continuous uplift of the region since the Quaternary is evidence of a long-term process with a distinctly episodic character. This might explain the absence of historic seismic activity in the area (Eyidoğan *et al.*, 1985).

Plotnikova *et al.* (1990) and Plotnikova *et al.* (1996) presented details regarding the production patterns and the gas pressure in the field, which indicate large temporal and spatial variations in pressure. Generally, injection patterns at Gazli were selective and erratic and resulted in drastic surface level changes of the gas-water contact, which rose by over 40 m in the northern and eastern parts of the field. Plotnikova *et al.* (1990) also reported subsidence of 16 mm in that area over the same time period 1972-1974. From these observations, Plotnikova *et al.* (1990) concluded that gas field operations impacted the seismic activity at Gazli. More precisely, the authors claimed that the events in 1976 were induced, while the second in 1984 was not. They based this distinction primarily on a statistical analysis of moderately sized seismicity prior to the two events.

In order to investigate seismic activity at Gazli in greater detail, a collaborative field survey was initiated in 1991 between the University of Grenoble and the Academy of Sciences of the Uzbek Republic (Bossu and Grasso, 1996; Bossu *et al.*, 1996). Bossu *et al.* (1996) found that the seismicity had migrated about 90 km unilaterally in the N240°E direction since the onset of the sequence in 1976. They interpreted this finding to indicate that an immature fault zone was present in the study area. This fault zone could explain both the major earthquakes and the variety of fault plane solutions computed through the field survey. See Figs 26 and 27 for details on the seismicity recorded during the field survey in 1991.

Bossu (1996) also came to the conclusion that the events at Gazli are most likely of tectonic origin. Bossu (1996) tested the possibility that crustal unloading might impact the occurrence of seismic activity at Gazli, as suggested by McGarr (1991) in the context of the California events. However, since water injection and gas extraction counteract, the stress disturbance created through that effect would only lie in the range of 0.06 bars and thus be comparable to tidal stresses.

Overall, despite the striking correlation between seismic activity and hydrocarbon production, the causal relation between the Gazli earthquake sequence and hydrocarbon production remains contentious. We stress that a substantial body of literature (see references in Adushkin *et al.* (2000)) on Gazli field was published in Russian and was not included in this review. In this context, it is interesting to note that Adushkin *et al.* (2000) presented Gazli as a case study for how an erratic production pattern might induce seismicity in hydrocarbon fields. In fact, they argued that the earthquake sequence might have been prevented through a more careful depletion strategy.

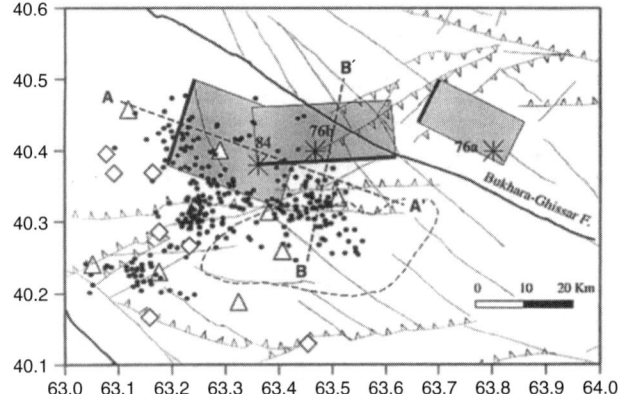

FIG. 26. Seismicity map of the 301 events registered and localized during the 1991 field survey carried out collaboratively by the University of Grenoble and the Academy of Sciences of the Uzbek Republic. The seismic network consisted of digital CEIS Lithoscope seismographs (triangles) and seven MEQ-800 seismographs (diamonds). The epicenters of the three main events (2, 4, and 10 in Table 2) are shown as stars. The geodetically determined fault planes (shaded rectangles), extent of the gas field (dashed line) and structural features are the same as in Fig. 25. The mapview of two cross sections AA′ and BB′ is also shown. The corresponding cross sections are shown in Fig. 27. *Source:* Figure reproduced from Bossu et al. (1996, Fig. 2, p. 961).

7. Conclusions and Outlook

Over the last few decades, the possibility that hydrocarbon production may induce earthquakes has become a dynamic field of research. The interest in this phenomenon was initially provoked by several prominent cases (e.g. Wilmington oil field or the earthquake series in the vicinity of Coalinga; see Section 6), in which a connection between seismicity and hydrocarbon production was suspected. Since then, a causal relationship between seismicity and hydrocarbon production has been suggested for 70 cases (see Table 1) based on spatial and temporal association of seismicity and field operation. Quantitative models (Sections 4 and 5) assess the change in stress or pore pressure as a consequence of production. However, these models are often hindered by a lack of detailed information about production patterns and by the inherent difficulty of judging whether a perturbation of a certain size is significant in comparison to the preexisting tectonic stress field or not. This difficulty highlights the challenge of drawing the line between natural and induced events.

Over the following years, induced seismicity is likely to remain an important research question, not only in the context of hydrocarbon production. Evaluating the possibility of inducing an event of moderate to large magnitude is not only relevant in the context of oil and gas reservoirs, but also for geothermal fields (Majer and Peterson, 2007), waste disposal wells (Nicholson *et al.*, 1988; Seeber *et al.*, 2004), and alternative usage of old reservoirs such as carbon sequestration. Continuing research targeted at assessing the seismic hazard associated with these operations is essential in order to perform them safely. In regions where induced seismicity is common, it has

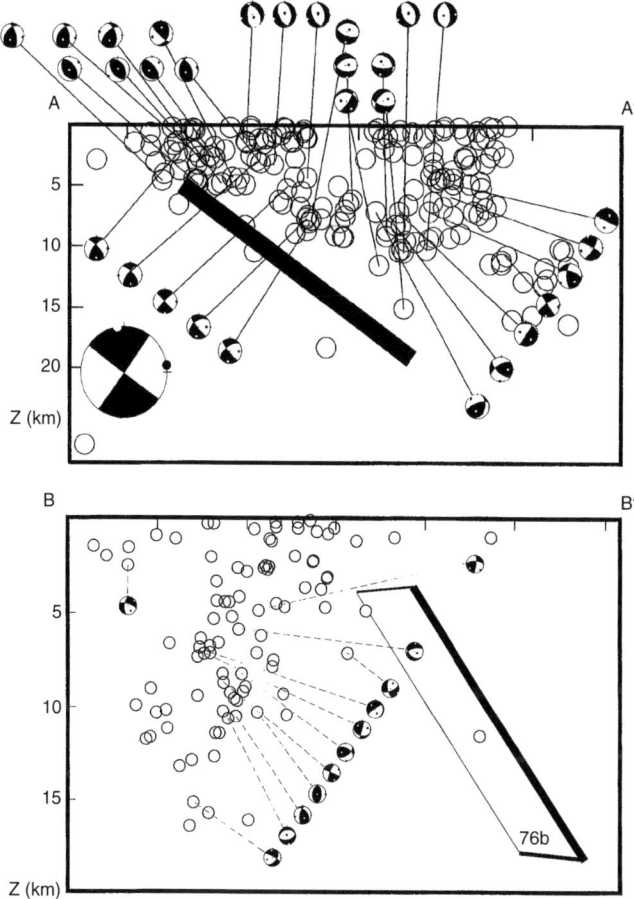

FIG. 27. Top: Cross section AA' along the geodetically determined fault plane for the 1984 event (for details on this event see Table 2 (event 10) and for the mapview of the cross section see Fig. 26). The focal mechanism of the 1984 event is enlarged in the left corner. Hypocenters and available focal mechanisms are shown for events registered during the 1991 field survey and located within 10 km from the vertical plane. The deepest event occurred at 26 km. Bottom: Corresponding hypocenters and focal mechanisms for cross section BB' (Fig. 26) in the vicinity of the fault plane for event 4 (Table 2). This time all events within 8 km distance were selected. The hypocenters recorded during the field survey define a south dipping zone contrary to the 1976b fault plane which is dipping to the north and does not seem to be associated with active seismicity.
Source: Figure reproduced from Bossu et al. (1996, Fig. 2, p. 961).

already become a social concern, particularly in the northern part of the Netherlands (Gussinklo *et al.*, 2001). As a reflection of that, since 2003, Dutch mining laws require quantitative estimates of the likelihood of future seismic activity and the associated damage for every onshore field. Already, the traditional tools of probabilistic seismic hazard analysis have been extended to model hazard from induced seismicity in actively producing fields (Van Eck *et al.*, 2006). Deriving an estimate for expected future hazard prior to production, however, requires knowledge of the factors that control induced seismicity, which are little understood. Van Eijs *et al.* (2006) investigated the correlation between reservoir properties and induced seismicity in the Netherlands (see Section 2.3). Adushkin *et al.* (2000) attempted a similar comparison on a global scale. Unfortunately, their results were based on reservoir characteristics different from those of Van Eijs *et al.* (2006), but also tend to indicate that hydrocarbon fields with induced seismicity deviate from a random sample of 200 hydrocarbon fields worldwide (see Fig. 28). Continued monitoring of induced seismicity over the following decades will allow determining whether these differences are statistically significant.

Arguably, the availability of high-quality data is the single most important requirement for evaluating the correlation between hydrocarbon production and seismicity. As a reflection of this necessity, the industry increasingly devotes resources to improving the instrumentation of oil fields and the data analysis. One prominent example is Valhall field which is equipped with a dense onshore and offshore seismic array (see Section 3). Another area of active research is the acquisition of time-dependent data (commonly referred to as time-lapse or 4D data). Since their introduction in the 1980s (Greaves and Fulp, 1987; Britton *et al.*, 1983), 4D techniques have advanced rapidly over the last few years. In 2001, Lumley (2001a) estimated that at that time there were about 75 active 4D seismic projects worldwide and the total annual expenditure on these projects ranged in the order of US $ 50-100 million. Most 4D monitoring projects are located in the North Sea. Of the total cumulative expenditure of US $ 500 million on 4D services, about $ 400 million (80%) has been spent in the North Sea (Lumley, 2004). Apart from mapping fluid flows, 4D images can potentially provide a better image of reservoir compartmentalization and fault properties, which could be useful for the optimal design of production facilities in complex reservoirs (Lumley, 2001a; Albright *et al.*, 1994). An example of a successful application of 4D seismic imaging for identifying the flow patterns at Gullfaks reservoir, North Sea, is shown in Fig. 29. The negative side of 4D monitoring is, apart from the costs involved, the fact that not all reservoirs are ideal candidates, because a sufficient signal-to-noise ratio is required to resolve subtle changes in the seismic image. Although 4D techniques are mostly used on similar data sets, it is also possible to cross match different data sets (Hall *et al.*, 2005). It will be interesting to see to which degree different data sets can be integrated in order to constrain changes in reservoir properties over time.

The acquisition of more and better data opens up new possibilities for geomechanical modeling of production-related changes in the reservoir over time. One concern often related to that of induced seismicity in hydrocarbon fields is surface subsidence (e.g. Goose Creek, Texas, USA (Pratt and Johnson, 1926), Wilmington, California, USA (Kosloff *et al.*, 1980a,b), Groningen, Rotliegendes, Netherlands (Barends *et al.*, 1995), Po Valley, Italy (Cassiani and Zoccatelli, 2000), and Costa Oriental, Lake Maracaibo, Venezuela (Xu, 2002)). The conditions under which subsidence is related to seismicity, however, are not trivial and little understood. In some hydrocarbon fields, subsidence is considered to be largely aseismic (e.g. Belridge, California, USA), but the moderate event at Ekofisk, North Sea, Norway (Ottemöller *et al.*, 2005) – for which this was long thought to be the case – highlighted the uncertainty associated with such judgments. In yet another case, at Wilmington oil field, California, substantial subsidence was associated

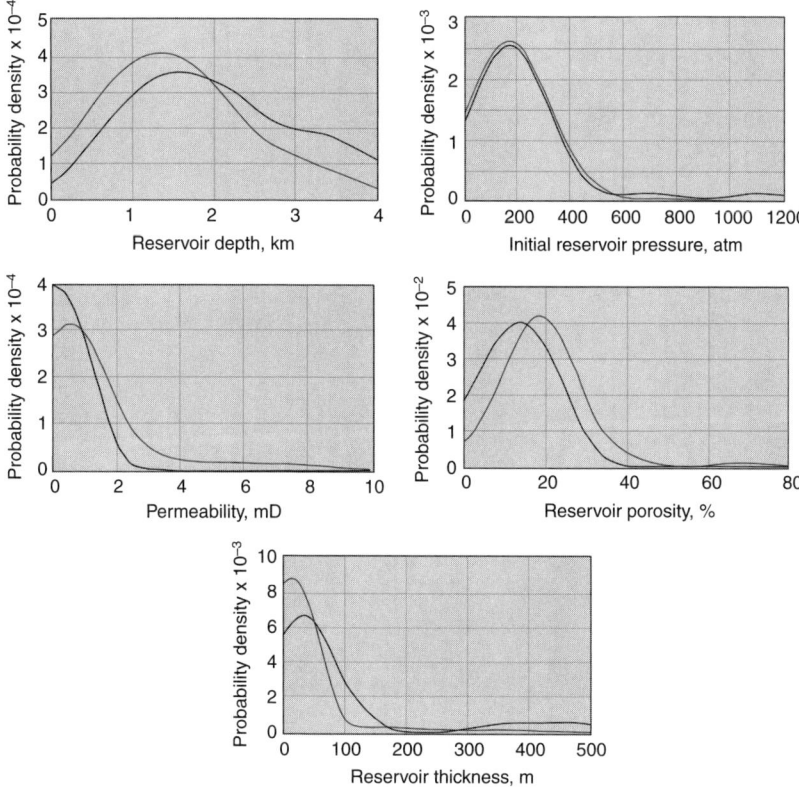

FIG. 28. Comparison of the properties of 40 hydrocarbon reservoirs exhibiting induced seismicity (black line) with a worldwide random sample of 200 reservoirs that do not (gray line). Shown are the probability distributions for the parameters, reservoir depth (km), initial pressure (atm), permeability (mD), porosity (%), and thickness (m).
© 2000, Schlumberger, Ltd. Used with permission.
Source: Figure reproduced from Adushkin et al. (2000, Fig. 7, p. 114).

with moderate seismic activity and the combination of both effects caused considerable damage in the port of Long Beach, California (Kosloff *et al.*, 1980a). The costs of the largest earthquake in the field on 17 November 1949 exceeded $ 9 million at that time (Nicholson and Wesson, 1990). These cases show that the correlation between the magnitude of subsidence and seismicity is complex. In fact, it is possible that potentially large stresses build up in a comparatively stiff rock, which due its stiffness will only yield modest subsidence. Finally, we note that in the case of Goose Creek Oil field, surface subsidence also had legal consequences. A brief description of the case can be found in Wang (2000). A more extensive description of the potential legal implications of induced seismicity in the USA was provided by Cypser and Davis (1998).

Recent efforts in geomechanical modeling have focused and will hopefully continue to focus on: (1) Capturing the changes in reservoir properties over time and verifying

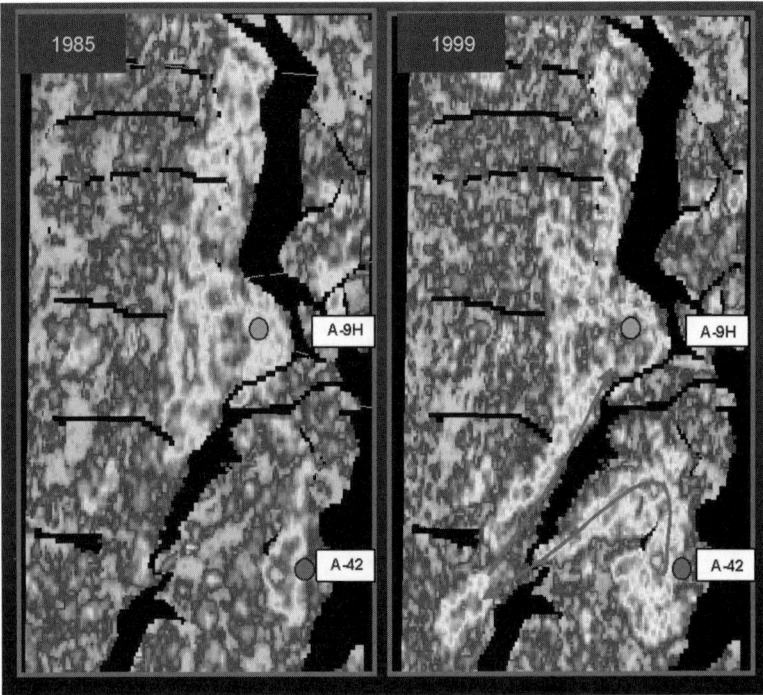

FIG. 29. Map of seismic amplitudes at Gullfaks reservoir in the North Sea, operated by Statoil. Two wells are indicated: WellA-9H is a production well and A-42 an injection well. The seismic amplitudes indicate that instead of flowing through the shortest possible path between the two wells, the injected gas takes a more convoluted route through the reservoir crossing a major fault boundary. This example demonstrates how time-lapse seismic imaging was successfully applied to identify fluid flow patterns.
Source: Figure reproduced from Lumley (2001b, Fig. 2, p. 1)

the modeling results with available data. An example highlighting the mutual benefits of high-quality data and sophisticated modeling is Ekofisk (Guilbot and Smith, 2002). Fig. 30 shows the map of the compaction distribution derived from seismic data (left) and the corresponding compaction map resulting from geomechanical modeling of the reservoir (right). (2) Analyzing the interaction of production, changes in reservoir properties, and pre-existing faults. This question has received particular attention in the Netherlands, because induced seismicity in hydrocarbon fields is thought to be predominantly related to the reactivation of pre-existing discontinuities (e.g. Mulders (2003)). (3) Making the connection between mechanical properties (e.g. reservoir compaction or surface subsidence), seismicity and stress corrosion or geochemical effects. At this point, the potential importance of these processes for subsidence has been recognized (e.g. at Ekofisk; see Section 4.3). It will be interesting to see whether they are relevant to earthquake generation as well. (4) The problem of nonlinear deformation and the connection to seismic activity. Although linear poroelastic models have been

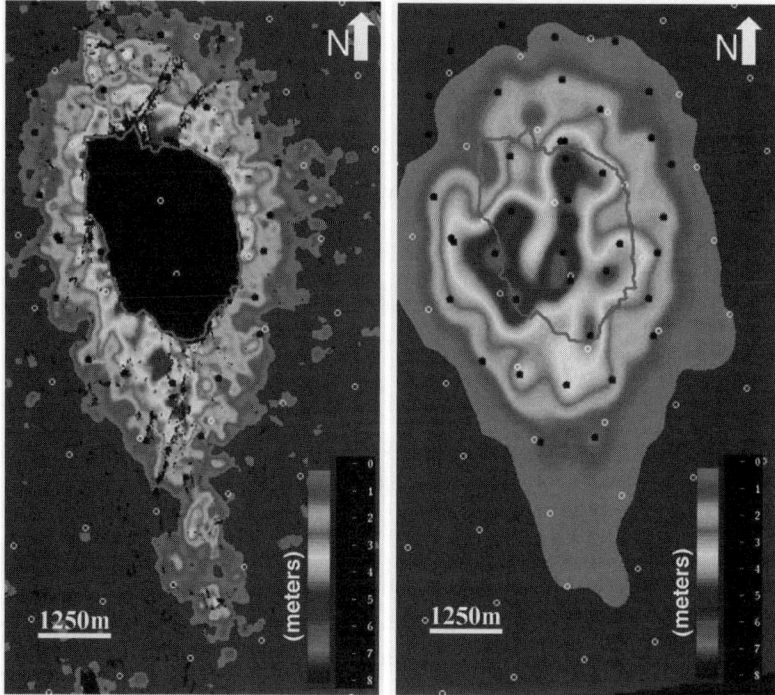

FIG. 30. Left: Compaction map of the reservoir at Ekofisk field derived from 4D seismic data. Right: Compaction map computed from geomechanical modeling. Water injection wells are displayed as black dots in both cases.
Source: Figure as given by Guilbot and Smith (2002, Fig. 8, p. 306 and Fig. 10, p. 307).

successfully applied to understanding induced seismicity (e.g. Segall *et al.* (1994)), the importance of nonlinear effects is evident from the mere magnitude of subsidence in hydrocarbon fields (Kosloff *et al.*, 1980b). One possible way forward is to differentiate between seismic events occurring within the reservoir materials as opposed to those occurring above and below. Assuming that anelastic effects are limited to the reservoir materials, one can couple a nonlinear model of the reservoir with an elastic model of the surrounding rock using Eshelby inclusion arguments (Segall and Fitzgerald, 1998).

Concluding, we argue that three of the main challenges in improving our understanding of induced seismicity in hydrocarbon fields will be (1) the development of tools for probabilistic seismic hazard assessments both prior to and during production, (2) the acquisition of high-quality data sets and the integration of different data sets to form a more complete picture of how the reservoir properties change over time, and (3) the development of more sophisticated geomechanical models. Finally, we believe that an important opportunity in the field is to develop a fruitful and active dialogue among industry, academia, and the public. The compilation of observational evidence and modeling advances presented in this study is intended to add to this dialogue. Hopefully, it will be complemented by and expanded upon by future reviews.

Acknowledgments

The idea and first draft of this review originated from a class project done for EPS-263, Earthquake Source Processes, at Harvard University. I am deeply indebted to Renata Dmowska and Jim Rice for encouragement and advice. I would also like to thank Paul Segall and two anonymous reviewers for valuable comments that have greatly improved the quality of the article. Finally, I thank Michael Fehler, Bradford H. Hager, Shawn Maxwell, and Nafi Toksöz for insightful discussions and Joe Hankins and Chris Sherratt from MIT libraries for their invaluable help.

References

Adushkin, V.V., Rodionov, V.N., Turuntaev, S., Yudin, A.E. (2000). Seismicity in the oil field. *Oilfield Rev.* **12** (2), 2–17.
Albright, J., Cassell, B., Dangerfield, J., Deflandre, J.-P., Johnstad, S., Withers, R. (1994). Seismic surveillance for monitoring reservoir changes. *Oilfield Rev.* **6** (1), 4–14.
Albright, J.N., Pearson, C.F. (1982). Acoustic emissions as a tool for hydraulic fracture location: Experience at the Fenton Hill hot dry rock site. *Soc. Petro. Eng. J.* **22** (4), 523–530.
Amorèse, D., Grasso, J.R. (1996). Rupture planes of the Gazli earthquakes deduced from local stress tensor calculation and geodetic data inversion: Geotectonic implications. *J. Geophys. Res.* **101** (B5), 11263–11274.
Amorèse, D., Grasso, J.R., Plotnikova, L.M., Nurtaev, B.S., Bossu, R. (1995). Rupture kinematics of the three gazli major earthquakes from vertical and horizontal displacement data. *Bull. Seism. Soc. Am.* **85** (2), 552–559.
Arrowsmith, S.J., Eisner, L. (2006). A technique for identifying microseismic multiplets and application to the Valhall field, North Sea. *Geophys.* **71** (2), V31–V40.
Baker, M.R., Doser, D.I., Luo, M. (1991). Geologic and oil field controls on earthquakes in the War-Wink Field, Delaware Basin. *Bulletin - West Texas Geol. Soc.* **31** (1), 5–12.
Baker, M.R., Mason, D.R., Doser, D.I. (1989). Seismicity (1978–79) associated with the War-Wink gas field, Ward County, Texas. *Seism. Res. Lett.* **60** (1), 31.
Baranova, V., Mustaqeem, A., Bell, S. (1999). A model for induced seismicity caused by hydrocarbon production in the Western Canada Sedimentary Basin. *Can. J. Earth Sc.* **36** (1), 47–64.
Bardainne, T., Gaillot, P., Dubos-Salle, N., Blanco, J., Sénéchal, G. (2006). Characterization of seismic waveforms and classification of seismic events using chirplet atomic decomposition. Example from the Lacq gas field (Western Pyrenees, France). *Geophys. J. Int.* **166** (2), 699–718.
Barends, F.B.J., Brouwer, F.J.J., Schroder, F.H. (1995). Land Subsidence. A.A. Balkema, Rotterdam.
Bennett, J.H., Sherburne, R.W. (1983). The 1983 Coalinga [Fresno County], California Earthquakes. Spec. Publi. (SP066), Calf. Div. Mines Geol.
Biot, M.A. (1941). General theory of three-dimensional consolidation. *J. Appl. Phys.* **12** (2), 155–164.
Block, L.V., Cheng, C.H., Fehler, M.C., Phillips, W.S. (1994). Seismic imaging using microearthquakes induced by hydraulic fracturing. *Geophys.* **59** (1), 102–112.
Bossu, R. (1996). Etude de la sismicité intraplaque de la région de Gazli (Ouzbékistan) et la localisation de la déformation sismique. PhD Thesis. L'Université Joseph Fourier, Grenoble.
Bossu, R., Grasso, J.R. (1996). Stress analysis in the intraplate area of Gazli, Uzbekistan, from different sets of earthquake focal mechanisms. *J. Geophys. Res.* **101** (B8), 17645–17659.
Bossu, R., Grasso, J.R., Plotnikova, L.M., Nurtaev, B., Frechet, J., Moisy, M. (1996). Complexity of intracontinental seismic faultings: The Gazli, Uzbekistan, sequence. *Bull. Seism. Soc. Am.* **86** (4), 959–971.
Bou-Rabee, F., Nur, A. (2002). The 1993 M4.7 Kuwait earthquake: Induced by the burning of the oil fields. *Kuwait J. Sci. Eng.* **29** (2), 155–163.
Britton, M.W., Martin, W.L., Leibrecht, R.J., Harmon, R.A. (1983). The street ranch pilot test of fracture-assisted steamflood technology. *J. Petr. Tech.* **35** (3), 511–522.

Calloi, P., Depanfilis, M., DiFilippo, D., Marcelli, I., Spadea, M.C. (1956). Terremoti della val Padana del 15–16 Maggio 1951. *Ann. Geofis.* **9**, 63–105.
Carslaw, A.R., Jaeger, C.J. (1959). Conduction of Heat in Solids. Clarendon Press, London.
Cassiani, G., Zoccatelli, C. (2000). Subsidence risk in Venice and nearby areas, Italy, owing to offshore gas fields: A stochastic analysis. *Environ. Eng. Geosc.* **6** (2), 115–128.
Chan, A.W., Zoback, M.D., Finkbeiner, T., Zinke, J. (2002). Production induced faulting and fault leakage in normal faulting regions; examples from the North Sea and Gulf of Mexico. In: AAPG annual convention with SEPM, Houston, TX, United States.
Console, R., Rosini, R. (1998). Non-double-couple microearthquakes in the geothermal fields of Larderello, central Italy. *Tectonophys.* **289** (1–3), 203–220.
Cook, C.C., Andersen, M.A., Halle, G., Gislefoss, E., Bowen, G.R. (2001). An approach to simulating the effects of water-induced compaction in a North Sea reservoir. *SPE Res. Eva. Eng.* **4** (2), 121–127.
Crampin, S., Peacock, S. (2005). A review of shear-wave splitting in the compliant crack-critical anisotropic Earth. *Wave Motion* **41** (1), 59–77.
Cypser, D.A., Davis, S.D. (1998). Induced seismicity and the potential for liability under U.S. law. *Tectonophys.* **289** (1–3), 239–255.
Dahm, T., Krüger, F., Stammler, K., Klinge, K., Kind, R., Wylegalla, K., Grasso, J.R. (2007). The 2004 M-w, 4.4 Rotenburg, Northern Germany, earthquake and its possible relationship with gas recovery. *Bull. Seism. Soc. Am.* **97** (3), 691–704.
Davis, S.D., Frohlich, C. (1993). Did (or will) fluid injection cause earthquakes: Criteria for a rational assessment. *Seism. Res. Lett.* **64** (3–4), 207–223.
Davis, S.D., Pennington, W.D. (1989). Induced seismic deformation in the Codgell oil-field of West Texas. *Bull. Seism. Soc. Am.* **79** (5), 1477–1494.
Davis, S.D., Nyffenegger, P.A., Frohlich, C. (1995). The 9 April 1993 Earthquake in South-Central Texas: Was it induced by fluid withdrawal?. *Bull. Seism. Soc. Am.* **85** (6), 1888–1895.
Davis, T.L., Namson, J., Yerkes, R.F. (1989). A cross-section of the Los-Angeles Area — Seismically active fold and thrust belt, the 1987 Whittier-Narrows Earthquake, and earthquake hazard. *J. Geophys. Res.* **94** (B7), 9644–9664.
Doser, D.I., Baker, M.R., Mason, D.B. (1991). Seismicity in the War-Wink Gas Field, Delaware Basin, West Texas, and its relationship to petroleum production. *Bull. Seism. Soc. Am.* **81** (3), 971–986.
Doser, D.I., Baker, M.R., Luo, M., Marroquin, P., Ballesteros, L., Kingwell, J., Diaz, H.L., Kaip, G. (1992). The Not So Simple relationship between seismicity and oil production in the Permian Basin, West Texas. *Pure Appl. Geophys.* **139** (3–4), 481–506.
Dost, B., Haak, H.W. (2008). Seismicity. In: Wong, T.E., Batjes, D.A.J., de Jager, J. (Eds.), Geology of the Netherlands. Edita-the Publishing House of the Royal Netherlands Academy of Arts and Sciences.
Dumas, D.B. (1979). Active seismic focus near Snyder, Texas. *Bull. Seism. Soc. Am.* **69** (4), 1295–1299.
Eaton, J.P. (1990). The Earthquake and its aftershocks from May 2 through September 30, 1983. In: The Coalinga, California, Earthquake of May 2, 1983. Vol. 1487, U.S. Geol. Survey Prof. Paper, pp. 113–170.
Ekström, G., Stein, R.S., Eaton, J.P., Eberhartphillips, D. (1992). Seismicity and geometry of a 110 km-long blind thrust-fault. 1. The 1985 Kettleman Hills, California, Earthquake. *J. Geophys. Res.* **97** (B4), 4843–4864.
Evans, D.G., Steeples, D.W. (1987). Microearthquakes near the Sleepy Hollow oil-field, Southwestern Nebraska. *Bull. Seism. Soc. Am.* **77** (1), 132–140.
Eyidoğan, H., Nábělek, J., Toksöz, M.N. (1985). The Gazli, USSR, 19 March 1984 Earthquake: The mechanism and tectonic implications. *Bull. Seism. Soc. Am.* **75** (3), 661–675.
Fehler, M., Jupe, A., Asanuma, H. (2001). More Than Cloud: New techniques for characterizing reservoir structure using induced seismicity. *Lead. Edge* **20** (3), 324–328.
Feignier, B., Grasso, J.R. (1990). Seismicity Induced by a Gas-Production: I. correlation of focal mechanisms and dome structure. *Pure Appl. Geophys.* **134** (3), 405–426.
Foulger, G.R., Julian, B.R., Hill, D.P., Pitt, A.M., Malin, P.E., Shalev, E. (2004). Non-double-couple microearthquakes at Long Valley caldera, California, provide evidence for hydraulic fracturing. *J. Volcanol. Geotherm. Res.* **132** (1), 45–71.
Geertsma, J. (1966). Problems of rock mechanics in petroleum production engineering. In: Proceed. 1st Cong. Int. Soc. Rock Mech., Vol. 1, Lisbon, pp. 585–594.

Geertsma, J. (1973). Land subsidence above compacting oil and gas reservoirs. *J. Petrol. Tech.* **25**, 734–744.
Gibbs, J.F., Healy, J.H., Raleigh, C.B., Coakley, J. (1973). Seismicity in the Rangely, Colorado, Area: 1962–1970. *Bull. Seism. Soc. Am.* **63** (5), 1557–1570.
Gibowicz, S.J., Lasocki, S. (2001). Seismicity induced by mining: Ten years later. *Adv. Geophys.* **44**, 39–180.
Gomberg, J., Wolf, L. (1999). Possible cause for an improbable earthquake: The 1997 Mw 4.9 southern Alabama earthquake and hydrocarbon recovery. *Geology* **27** (4), 367–370.
Goodier, J.N. (1937). On the integration of the thermo–elastic equations. *Phil. Mag.* **7**, 1017–1032.
Gough, D.I., Gough, W.I. (1970). Load-induced earthquakes at Lake Kariba—II. *Geophys. J. Int.* **21** (1), 79–101.
Grasso, J.R. (1992a). Effects of the gas extraction on the seismic and aseismic slips in the neighborhood of meillon-st-faust gas field. Technical report. Report to EIF-Aquitaine Company.
Grasso, J.R. (1992b). Mechanics of seismic instabilities induced by the recovery of hydrocarbons. *Pure Appl. Geophys.* **139** (3–4), 507–534.
Grasso, J.R., Feignier, B. (1990). Seismicity induced by Gas-Production: 2. Lithology correlated events, induced stresses and deformation. *Pure Appl. Geophys.* **134** (3), 427–450.
Grasso, J.R., Sornette, D. (1998). Testing self-organized criticality by induced seismicity. *J. Geophys. Res.* **103** (B12), 29965–29987.
Grasso, J.R., Wittlinger, G. (1990). 10 years of seismic monitoring over a gas field area. *Bull. Seism. Soc. Am.* **80** (2), 450–473.
Grasso, J.R., Gratier, J.P., Gamond, J.F., Paumier, J.C. (1992). Stress transfer and seismic instabilities in the upper crust — Example of the Western Pyrenées. *J. Struct. Geol.* **14** (8–9), 915–924.
Greaves, R.J., Fulp, T.J. (1987). 3-dimensional seismic monitoring of an enhanced oil-recovery process. *Geophys.* **52** (9), 1175–1187.
Guha, S. (2000). Induced Earthquakes. Kluwer Academic Publishers.
Guha, S.K., Gosavi, P.D., Padale, J.G., Marwadi, S.C. (1971). An earthquake cluster at Koyna. *Bull. Seism. Soc. Am.* **61** (2), 297–315.
Guilbot, J., Smith, B. (2002). 4-D constrained depth conversion for reservoir compaction estimation: Application to Ekofisk Field. *Lead. Edge* **21** (3), 302–308.
Gupta, H.K., Rastogi, B.K. (1976). Developments in Geotechnical Engineering 11. Elsevier Scientific Publishing Co., Amsterdam, (chapter Dams and Earthquakes), p. 299.
Gussinklo, H.J., Haak, H.W., Quadvlieg, R.C.H., Schutjens, P.M.F.M., Vogelaar, L. (2001). Subsidence, tremors and society. *Geologie en Mijnbouw. Netherl. J. Geosci.* **80** (1), 121–136.
Guyoton, F., Grasso, J.R., Volant, P. (1992). Interrelation between induced seismic instabilities and complex geologic structure. *Geophys. Res. Lett.* **19** (7), 705–708.
Haak, H.W. (1991). Seismische Analyse van der Aardbeving bij Emmen op 15 februari 1991. In 'Koninklijk Nederlands Meteorologisch Instituut'. Ministerie van Verkeer en Waterstaat, p. 14.
Hall, S.A., Kendall, J.-M. (2003). Fracture characterization at Valhall: Application of P-wave amplitude variation with offset and azimuth (AVOA) analysis to a 3D ocean-bottom data set. *Geophys.* **68** (4), 1150–1160.
Hall, S.A., MacBeth, C., Barkved, O.I., Wild, P. (2005). Cross-matching with interpreted warping of 3D streamer and 3D ocean-bottom-cable data at Valhall for time-lapse assessment. *Geophys. Prosp.* **53** (2), 283–297.
Harding, S.T. (1981). Induced seismic Cogdell Canyon Reef oil field. Technical report. U.S. Geol. Surv., Open File Rept. 81–167.
Hauksson, E., Jones, L.M., Davis, T.L., Hutton, L.K., Brady, A.G., Reasenberg, P.A., Michael, A.J., Yerkes, R.F., Williams, P., Reagor, G., Stover, C.W., Bent, A.L., Shakal, A.K., Etheredge, E., Porcella, R.L., Bufe, C.G., Johnston, M.J.S., Cranswick, E. (1988). The 1987 Whittier Narrows Earthquake in the Los-Angeles Metropolitan Area, California. *Science* **239** (4846), 1409–1412.
Heggheim, T., Madland, M.V., Risnes, R., Austad, T. (2005). A chemical induced enhanced weakening of chalk by seawater. *J. Petrol. Sci. Eng.* **46** (3), 171–184.
Helbig, K., Thomsen, L. (2005). 75th Anniversary Paper - 75-plus years of anisotropy in exploration and reservoir seismics: A historical review of concepts and methods. *Geophys.* **70** (6), 9ND–23ND.

Hermansen, H., Landa, G.H., Sylte, J.E., Thomas, L.K. (2000). Experiences after 10 years of waterflooding the Ekofisk Field, Norway. *J. Petrol. Sci. Eng.* **26** (1–4), 11–18.

Herquel, G., Wittlinger, G. (1994). Anisotropy at the Lacq hydrocarbon field (France) from shear-wave splitting. *Geophys. Res. Lett.* **21** (24), 2621–2624.

Horner, R.B., Barclay, J.E., MacRae, J.M. (1994). Earthquakes and hydrocarbon production in the Fort St. John area of northeastern British Columbia. *Can. J. Expl. Geophys.* **30** (1), 39–50.

Jones, R.H., Stewart, R.C. (1997). A method for determining significant structures in a cloud of earthquakes. *J. Geophys. Res.* **102** (B4), 8245–8254.

Jupe, A., Cowles, J., Jones, R. (1998). Microseismic monitoring: Listen and see the reservoir. *World Oil* **219** (12), 171–174.

Kanamori, H., Hauksson, E. (1992). A slow earthquake in the Santa-Maria Basin, California. *Bull. Seism. Soc. Am.* **82** (5), 2087–2096.

Key, S.C., Pederson, S.H., Smith, B.A. (1998). Adding value to reservoir management with seismic monitoring technologies. *Lead. Edge* **17** (4), 515–519.

Kisslinger, C. (1976). Theories of mechanisms of induced seismicity. *Eng. Geology* **10** (2–4), 85–98.

Kosloff, D., Scott, R.F., Scranton, J. (1980a). Finite element simulation of wilmington oil field subsidence: I. linear modeling. *Tectonophys.* **65** (3–4), 339–368.

Kosloff, D., Scott, R.F., Scranton, J. (1980b). Finite element simulation of wilmington oil field subsidence: II. nonlinear modeling. *Tectonophys.* **70** (1–2), 159–183.

Kouznetsov, O., Sidorov, V., Katz, S., Chilingarian, G. (1994). Interrelationships among seismic and short-term tectonic activity, oil and gas production, and gas migration to the surface. *J. Petrol. Sci. Eng.* **13** (1), 57–63.

Kovach, R.L. (1974). Source mechanisms for wilmington oil field, California, subsidence earthquakes. *Bull. Seism. Soc. Am.* **64** (3), 699–711.

Kreitler, C. (1978). Faulting and land subsidence from ground-water and hydrocarbon production, Houston-Galveston, Texas. Technical report. Bureau of Economic Geology, University of Texas at Austin, Report 8, 22 pp.

Kristiansen, T.G., Barkved, O., Patillo, P.D. (2000). Use of passive seismic monitoring in well and casing design in the compacting and subsiding valhall field, North Sea. *Soc. Petr. Eng.* SPE, 65134.

Lahaie, F., Grasso, J.R. (1999). Loading rate impact on fracturing pattern: Lessons from hydrocarbon recovery, Lacq gas field, France. *J. Geophys. Res.* **104** (B8), 17,941–17,954.

Love, A. (1944). A Treatise on the Mathematical Theory of Elasticity. Dover, New York.

Lumley, D.E. (2001a). Time-lapse seismic reservoir monitoring. *Geophys.* **66** (1), 50–53.

Lumley, D.E. (2001b). The next wave in reservoir monitoring: The instrumented oil field. *Lead. Edge* **20** (6), 640–648.

Lumley, D.E. (2004). Business and technology challenges for 4D seismic reservoir monitoring. *Lead. Edge* **23** (11), 1166–1168.

Majer, E.L., Peterson, J.E. (2007). The impact of injection on seismicity at The Geysers, California Geothermal Field. *Int. J. Rock Mech. & Min. Sci.* **44** (8), 1079–1090.

Marroouin, P., Ballestros, L., Doser, D.I., Baker, M.R. (1992). Earthquakes associated with the Keystone oil field, West Texas. *Seism. Res. Lett.* **63** (1), 19.

Martin, R.J. (1972). Time-dependent crack growth in quartz and its application to the creep of rocks. *J. Geophys. Res.* **77** (8), 1406–1419.

Maury, V.M.R., Grasso, J.R., Wittlinger, G. (1992). Monitoring of subsidence and induced seismicity in the Lacq gas field (France): The consequences on gas production and field operation. *Eng. Geol.* **32** (3), 123–135.

Maxwell, S.C., Urbancic, T.I. (2005). The potential role of passive seismic monitoring for real-time 4D reservoir characterization. *SPE Res. Ev. Eng. (SPE 89071)* **8** (1), 70–76.

McGarr, A. (1991). On a possible connection between three major earthquakes in California and oil production. *Bull. Seism. Soc. Am.* **81** (3), 948–970.

McGarr, A., Simpson, D. (1997). Keynote lecture: A broad look at induced and triggered seismicity. In: Gibowicz, S.J., Lasocki, S. (Eds.), Rockbursts and Seismicity in Mines. Balkema Rotterdam, pp. 385–396.

Mereu, R.F., Brunet, J., Morrissev, K., Price, B., Yapp, A. (1986). A study of the microearthquakes of the gobles oil field area of Southwestern Ontario. *Bull. Seism. Soc. Am.* **76** (5), 1215–1223.

Miaoyueh, W., Maoyuan, Y., Yuliang, H., Tzuchiang, L., Yuntai, C., Yen, C., Jui, F. (1976). Preliminary study on mechanism of reservoir impounding Earthquakes at Hsinfengkiang. *Scientia Sinica* **19** (1), 149–169.
Milne, W.G. (1970). The Snipe Lake, Alberta earthquake of March 8, 1970. *Can. J. Earth Sci.* **7** (6), 1564–1567.
Mindlin, R.D., Cheng, D.H. (1950). Thermoelastic stress in the semi-infinite solid. *J. Appl. Phys* **21** (9), 931–933.
Mulders, F.M.M. (2003). Modelling of stress development and fault slip in and around a producing gas reservoir. PhD thesis. TU Delft, Netherlands.
Murria, J. (1997). Earthquake geotechnical engineering aspects of the protection dikes of the Costa Oriental of Lake Maracaibo, Venezuela. In: Venezuelan Foundation for Seismological Research, Caracas, Venezuela.
Nagel, N.B. (2001). Compaction and subsidence issues within the petroleum industry: From wilmington to ekofisk and beyond. *Phys. Chem. Earth, Part A* **26** (1–2), 3–14.
Nagelhout, A.C.G., Roest, J.P.A. (1997). Investigating fault slip in a model of an underground gas storage facility. *Int. J. Rock Mech. & Min. Sci.* **34** (3–4), Paper No. 212.
Nicholson, C., Wesson, R.L. (1990). Earthquake hazard associated with deep well injections a report of the U.S. /Environmental Protection Agency. Technical report. U.S. Geol. Surv. Bull. 1951.
Nicholson, C., Wesson, R.L. (1992). Triggered earthquakes and deep well activities. *Pure Appl. Geophys.* **139**, 561–578.
Nicholson, C., Roeloffs, E., Wesson, R.L. (1988). The Northeastern Ohio Earthquake of 31 January 1986: Was it induced? *Bull. Seism. Soc. Am.* **78** (1), 188–217.
Odonne, F., Ménard, I., Massonnat, G.J., Rolando, J.P. (1999). Abnormal reverse faulting above a depleting reservoir. *Geology* **27** (2), 111–114.
Olofsson, B., Probert, T., Kommedal, J.H., Barkved, O.I. (2003). Azimuthal anisotropy from the Valhall 4D 3D survey. *Lead. Edge* **22** (12), 1228–1235.
Orr, C.D., Keller, G.R. (1981). Keystone Field, Winkler County, Texas; an examination of seismic activity, in-situ stresses, effective stresses, and secondary recovery. *Earthq. Notes* **52** (1), 29–30.
Ottemöller, L., Nielsen, H.H., Atakan, K., Braunmiller, J., Havskov, J. (2005). The 7 May 2001 induced seismic event in the Ekofisk oil field, North Sea. *J. Geophys. Res.* **110** (B10), B10301.
Ovens, J.E.V., Larsen, F.P., Cowie, D.R. (1997). Making sense of water injection fractures in the Dan Field. *Soc. Petr. Eng. AIME* **1997**, 887–901.
Pearson, C. (1981). The relationship between microseismicity and high pore pressures during hydraulic stimulation experiments in low permeability granitic rocks. *J. Geophys. Res.* **86** (NB9), 7855–7864.
Pennington, W.D., Davis, S.D., Carlson, S.M., Dupree, J., Ewing, T.E. (1986). The evolution of seismic barriers and asperities caused by the depressuring of fault planes in oil and gas of South Texas. *Bull. Seism. Soc. Am.* **76** (4), 939–948.
Perrodon, A. (1983). Dynamics of oil and gas accumulations. Elf Aquitaine.
Phillips, W.S., Fairbanks, T.D., Rutledge, J.T., Anderson, D.W. (1998). Induced microearthquake patterns and oil-producing fracture systems in the Austin chalk. *Tectonophys.* **289** (1–3), 153–169.
Plotnikova, I.M., Nurtaev, B.S., Grasso, J.R., Matasova, L.M., Bossu, R. (1996). The character and extent of seismic deformation in the focal zone of Gazli earthquakes of 1976 and 1984, $M > 7.0$. *Pure Appl. Geophys.* **147** (2), 377–387.
Plotnikova, I.M., Flyonova, M.G., Machmudova, V.I. (1990). Induced seismicity in the Gazli gas field region. *Gerlands Beiträge zur Geophysik* **99**, 309–319.
Pratt, W.E., Johnson, D.W. (1926). Local subsidence of the Goose Creek oil field. *J. Geology* **34**, 577–590.
Raleigh, C.B., Healy, J.H., Bredehoeft, J.D. (1976). An experiment in Earthquake control at Rangely, Colorado. *Science* **191** (4233), 1230–1237.
Raleigh, C., Healy, J.H., Bredehoeft, J.D. (1972). Faulting and Crustal Stress at Rangely, Colorado. In: Flow and Fracture of Rocks. In: Geophys. Monogr. Am. Geophys. Union, Vol. 16, pp. 275–284.
Rice, J.R., Cleary, M.P. (1976). Some basic stress-diffusion solutions for fluid-saturated elastic porous media with compressible constituents. *Rev. Geophys. Space Phys.* **14**, 227–241.
Roest, J.P.A., Mulders, F.M.M. (2000). Overview modelling gas production-induced seismicity mechanisms. In: Proceedings of the EUROCK 2000 Symposium, Aachen, 27–31 March 2000, pp. 333–338.

Roest, J.P.A., Kuilman, W. (1994). Geomechanical analysis of small earthquakes at the Eleveld gas reservoir. In: 'Eurock '94; SPE/ISRM international conference, Delft, Netherlands', pp. 573–580.
Rothe, G.H., Lui, C.Y. (1983). Possibility of induced seismicity in the vicinity of the Sleepy Hollow oil field, Southwestern Nebraska. *Bull. Seism. Soc. Am.* **73** (5), 1357–1367.
Rutledge, J.T., Phillips, W.S. (2003). Hydraulic stimulation of natural fractures as revealed by induced microearthquakes, Carthage Cotton Valley gas field, east Texas. *Geophys.* **68** (2), 441–452.
Rutledge, J.T., Phillips, W.S., Schuessler, B.K. (1998). Reservoir characterization using oil-production-induced microseismicity, Clinton County, Kentucky. *Tectonophys.* **289** (1–3), 129–152.
Rymer, M.J., Ellsworth, W.L. (Eds.) (1990). The Coalinga, California, Earthquake of May 2, 1983. U.S. Geol. Survey Prof. Paper 1487, U.S. Government Printing Office, Washington.
Scholz, C.H. (1972). Static fatigue of quartz. *J. Geophys. Res.* **77** (11), 2104–2114.
Seeber, L., Armbruster, J.G., Kim, W.Y. (2004). A fluid-injection-triggered earthquake sequence in Ashtabula, Ohio: Implications for seismogenesis in stable continental regions. *Bull. Seism. Soc. Am.* **94** (1), 76–87.
Segall, P. (1985). Stress and Subsidence resulting from subsurface fluid withdrawal in the epicentral region of the 1983 Coalinga Earthquake. *J. Geophys. Res.* **90** (B8), 6801–6816.
Segall, P. (1989). Earthquakes triggered by fluid extraction. *Geology* **17** (10), 942–946.
Segall, P. (1992). Induced stresses due to fluid extraction from axisymmetric reservoirs. *Pure Appl. Geophys.* **139** (3–4), 535–560.
Segall, P., Yerkes, R.F. (1990). Stress and fluid-pressure changes associated with oil-field operations: A critical assessment of effects in the focal region of the earthquake. In: The Coalinga, California, Earthquake of May 2, 1983. Vol. 1487. U.S. Geol. Survey Prof. Paper, pp. 259–272.
Segall, P., Fitzgerald, S.D. (1998). A note on induced stress changes in hydrocarbon and geothermal reservoirs. *Tectonophys.* **289** (1–3), 117–128.
Segall, P., Grasso, J.R., Mossop, A. (1994). Poroelastic stressing and induced seismicity near the lacq gas field, southwestern france. *J. Geophys. Res.* **99** (B8), 15423–15438.
Sheets, M.M. (1979). Oil fields and their relation to subsidence and active surface faulting in the Houston area. Houston Geol. Soc., Houston, Tex, USA.
Shen, C.K., Chen, H.C., Huang, L.S., Yang, C.Y., Chang, C.H., Li, T.C., Wang, T.C., Lo, H.H. (1974). Earthquakes induced by Reservoir Impounding and their effect on Hsinfengkiang Dam. *Scientia Sinica* **17** (2), 239–272.
Shouzhong, D., Huanpeng, Z., Aixiang, G. (1987). Rare seismic clusters induced by water injection in the Jiao well 07 in Shengli oil field. *Earthq. Res. China* **1** (2), 313–314.
Simpson, D.W. (1976). Seismicity changes associated with reservoir loading. *Eng. Geol.* **10** (2–4), 123–150.
Simpson, D.W., Leith, W. (1985). The 1976 and 1984 Gazli, USSR, earthquakes-were they induced? *Bull. Seism. Soc. Am.* **75** (5), 1465–1468.
Singh, D.D., Rastogi, B.K., Gupta, H.K. (1975). Surface-wave Radiation-pattern and Source Parameters of Koyna Earthquake of December 10, 1967. *Bull. Seism. Soc. Am.* **65** (3), 711–731.
Sminchak, J., Gupta, N. (2003). Aspects of induced seismic activity and deep-well sequestration of carbon dioxide. *Environ. Geosc.* **10** (2), 81–89.
Smirnova, M.N. (1968). Effects of earthquakes on the oil yield of the gudermes field (Northeastern Caucasus), Izvestia. *Earth Physics* **12**, 760–763.
Stein, S., Wiens, D.A., Fujita, K. (1982). The 1966 Kremasta reservoir earthquake sequence. *Earth Planet. Sci. Lett.* **59** (1), 49–60.
Sze, E. (2005). Induced seismicity analysis for reservoir characterization at a petroleum field in Oman. PhD Thesis. Massachusetts Institute of Technology.
Talebi, S., Boone, T.J. (1998). Source parameters of Injection-induced Microseismicity. *Pure Appl. Geophys.* **153** (1), 113–130.
Teufel, L.W., Rhett, D.W., Farrell, H.E. (1991). Effect of reservoir depletion and pore pressure drawdown on in situ stress and deformation in the Ekofisk Field, North Sea. In: J. C. Rogiers (Ed.). 'Rock Mechanics as a Multidisciplinary Science'. Balkema, Rotterdam. pp. 63–72.
Torsaeter, O. (1984). An experimental study of water imbibition in chalk from the Ekofisk Field. *Soc. Petrol. Eng. AIME*.

Turuntaev, S.B., Razumnaya, O.A. (2002). An application on induced seismicity data analysis for detection of spatial structures and temporal regimes of deformation processes in hydrocarbon fields. *Pure Appl. Geophys.* **159** (1–3), 421–447.

Van Dok, R.R., Gaiser, J.E., Byerley, G. (2003). Near-surface shear-wave birefringence in the North Sea: Ekofisk 2D/4C test. *Lead. Edge* **22** (12), 1236–1242.

Van Eck, T., Goutbeek, F., Haak, H., Dost, B. (2006). Seismic hazard due to small-magnitude, shallow-source, induced earthquakes in The Netherlands. *Eng. Geol.* **87** (1–2), 105–121.

Van Eijs, R.M.H.E., Mulders, F.M.M., Nepveu, M., Kenter, C.J., Scheffers, B.C. (2006). Correlation between hydrocarbon reservoir properties and induced seismicity in the Netherlands. *Eng. Geol.* **84** (3–4), 99–111.

Van Wees, J.D., Orlic, B., van Eijs, R., Zijl, W., Jongerius, P., Schreppers, G.J., Hendriks, M., Cornu, T. (2003). Integrated 3D geomechanical modelling for deep subsurface deformation; a case study of tectonic and human-induced deformation in the eastern Netherlands. *Geol. Soc. Spec. Pub.* **212**, 313–328.

Vavrycuk, V., Bohnhoff, M., Jechumtalova, Z., Kolar, P., Sileny, J. (2008). Non-double-couple mechanisms of microearthquakes induced during the 2000 injection experiment at the KTB site, Germany: A result of tensile faulting or anisotropy of a rock? *Tectonophys.* **456** (1–2), 74–93.

Wang, H. (2000). Theory of Linear Poroelasticity with Applications to Geomechanics and Hydrogeology. Princeton University Press.

Wang, M., Yang, M., Hu, Y., Li, T., Chen, Y., Yen, C., Jui, F. (1976). Mechanism of Reservoir Impounding Earthquakes at Hsinfengkian and a preliminary endeavor to discuss their cause. *Eng. Geol.* **10** (2–4), 331–351.

Wetmiller, R.J. (1986). Earthquakes near Rocky Mountain House, Alberta, and Relationship to Gas Production. *Can. J. Earth Sciences* **32** (2), 172–181.

Wiprut, D., Zoback, M.D. (2000). Fault reactivation and fluid flow along a previously dormant fault in the northern North Sea. *Geology* **28** (7), 595–598.

Xu, H. (2002). Production induced reservoir compaction and surface subsidence, with applications to 4D seismic. PhD Thesis. Stanford University.

Yerkes, R.F., Castle, R.O. (1976). Seismicity and faulting attributable to fluid extraction. *Eng. Geology* **10** (2–4), 151–167.

Yerkes, R.F., Levine, P., Wentworth, C.M. (1990). Abnormally high fluid pressures in the region of the Coalinga Earthquake sequence and their significance. In: The Coalinga, California, Earthquake of May 2, 1983. Vol. 1487. U.S. Geol. Survey Prof. Paper, pp. 235–257.

Zoback, M.D., Zinke, J.C. (2002). Production-induced normal faulting in the Valhall and Ekofisk Oil Fields. *Pure Appl. Geophys.* **159** (1–3), 403–420.

PHENOMENOLOGY OF TSUNAMIS: STATISTICAL PROPERTIES FROM GENERATION TO RUNUP

Eric L. Geist[1]

Abstract

Observations related to tsunami generation, propagation, and runup are reviewed and described in a phenomenological framework. In the three coastal regimes considered (near-field broadside, near-field oblique, and far field), the observed maximum wave amplitude is associated with different parts of the tsunami wavefield. The maximum amplitude in the near-field broadside regime is most often associated with the direct arrival from the source, whereas in the near-field oblique regime, the maximum amplitude is most often associated with the propagation of edge waves. In the far field, the maximum amplitude is most often caused by the interaction of the tsunami coda that develops during basin-wide propagation and the nearshore response, including the excitation of edge waves, shelf modes, and resonance. Statistical distributions that describe tsunami observations are also reviewed, both in terms of spatial distributions, such as coseismic slip on the fault plane and near-field runup, and temporal distributions, such as wave amplitudes in the far field. In each case, fundamental theories of tsunami physics are heuristically used to explain the observations.

KEYWORDS: Tsunami, Earthquake, Landslide, Phenomenology, Statistical, Runup, Tide gage.
© 2009 Elsevier Inc.

1. Introduction

Major tsunamis are often accompanied by some aspect of unexpected behavior, relative to a simple deterministic understanding of tsunami physics. Typically, the unexpected behavior is related to the amplitude, runup, and timing of the largest wave when a tsunami strikes the coast. Kanamori (1972) first classified earthquakes that result in tsunamis with greater than expected overall severity relative to earthquake magnitude as "tsunami earthquakes", examining specifically the highly destructive tsunamis emanating from the 1896 Sanriku and 1946 Aleutian earthquakes. Other examples of tsunami earthquakes that generated unexpectedly severe and devastating tsunamis relative to the magnitude of the generating earthquake include the 1992 Nicaragua event (Kanamori and Kikuchi, 1993) and the 1994 and 2006 Java events (Abercrombie et al., 2001; Polet and Thio, 2003; Ammon et al., 2006). It has been argued that a portion of the 2004 Sumatra-Andaman earthquake also had behavior similar to tsunami earthquakes (Seno and Hirata, 2007). Not only can the overall severity be unexpectedly large, but the runup along isolated portions of the coast can be unexpectedly large relative to the overall runup associated with the tsunami. Such a case occurred with

---

[1] Author thanks e-mail: egeist@usgs.gov

the 1993 Hokkaido tsunami having $M_w = 7.6$ (Mendoza and Fukuyama, 1996), in which a maximum runup value of 31.7 m was measured at a small valley that opens to a pocket beach (Shuto and Matsutomi, 1995), owing to local focusing of the wave by the nearshore bathymetry. Local high runup values can also be associated with strong variability in the source processes, as in the case for runup along the western Aceh coastline associated with the 2004 Sumatra-Andaman earthquake (Hirata et al., 2006; Fujii and Satake, 2007). Another type of unexpected behavior when the coastal hydrodynamic response is not taken into account is a delay in the expected timing of the maximum runup arrival. Crescent City, California is particularly prone to large tsunami arrivals occurring hours after the first arrival (Dengler et al., 2009). Earlier than expected tsunami arrivals have also been reported, as is the case for the 1983 Japan Sea tsunami (Shuto et al., 1995).

In an attempt to understand tsunamis better in terms of their observed behavior, I examine the link between geophysical phenomena that generate tsunamis and the ensuing tsunami waveforms from a phenomenological perspective (cf. Kanamori and Brodsky (2004)). The focus is on providing a review of tsunami observations, from generation through open-ocean propagation and runup. For the tsunami source, these observations include parameters, such as heterogeneous coseismic slip on the fault plane, that directly influence the initial conditions for tsunami propagation. I review the historical development of stochastic slip models that encompass much of the variability associated with the complex dynamics of fault rupture and that are constrained by the inversion of seismograms. For tsunami observations in the near field, I examine the relationship between source heterogeneity and spatial distribution of broadside runup and wave characteristics at oblique propagation paths. In addition, the timing of the maximum amplitude arrival is examined, both broadside and at oblique propagation paths from the source (Fig. 1). In the far field, I examine the statistics of wave amplitude as recorded by deep-ocean bottom pressure gages and coastal tide gages, that yield information on the origin of the stochastic nature of the tsunami wavefield. In each case, the description of tsunami event observations is heuristically guided by a review of analytic theory for the purpose of understanding various aspects of tsunami physics, ranging from the dynamic rupture of faults to nearshore wave behavior. Significant observations and their underlying physics are presented in this paper as several working hypotheses.

In accordance with the phenomenological objective of this study, results from numerical models are minimally used. The reason for this is to avoid the problem of needing high-resolution bathymetric and topographic data to accurately simulate complex hydrodynamics (e.g. nonlinearity, turbulence, rotational flow, etc.) in the nearshore region. The problem is most obvious when trying to simulate an entire tide gage record of a tsunami, where late arrivals may originate from coastline irregularities distant from the tide gage station. Because of the difficulty in collecting high-resolution nearshore bathymetry (especially in the surf zone) over long stretches of coastline, there are very few cases where late tsunami arrivals can be accurately predicted using numerical methods. Indeed, many of the late arrivals that make up the coda (i.e. the part of a tsunami time series that follows the first arrival) of a transoceanic tsunami originate from the fine-scale details of wave reflection from distant bathymetric features (e.g. see Koshimura et al. (2008); Kowalik et al. (2008)). Many of the same issues surrounding resolution in numerical ocean and coastal circulation models discussed by Greenberg et al. (2007) also apply to tsunami propagation models.

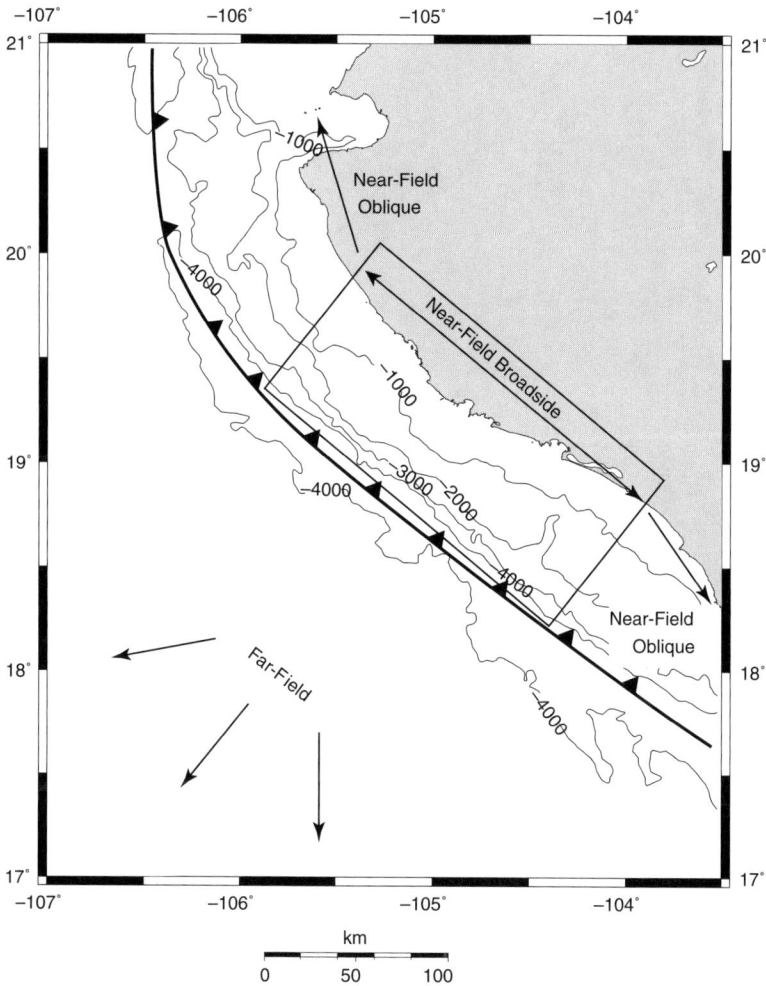

FIG. 1. Mapview of three different regimes for describing tsunami observations relative to projection of inter-plate thrust earthquake (solid rectangle): near-field broadside, near-field oblique, and far field (transoceanic).

2. General Description of Tsunami Generation Physics

Tsunamis are gravity waves initially induced by elevation changes in the water column commonly caused by various submarine geologic processes. The efficiency of solid-earth dynamic processes in transferring potential energy into the water column during tsunami generation is dependent on the time scale of geologic movements relative to the phase speed of long waves in the ocean. For the purpose of comparing rapid earthquake displacements with relatively slower landslide displacements, it is useful to mention the experimental and computational results of Hammack (1973). For the "impulsive"

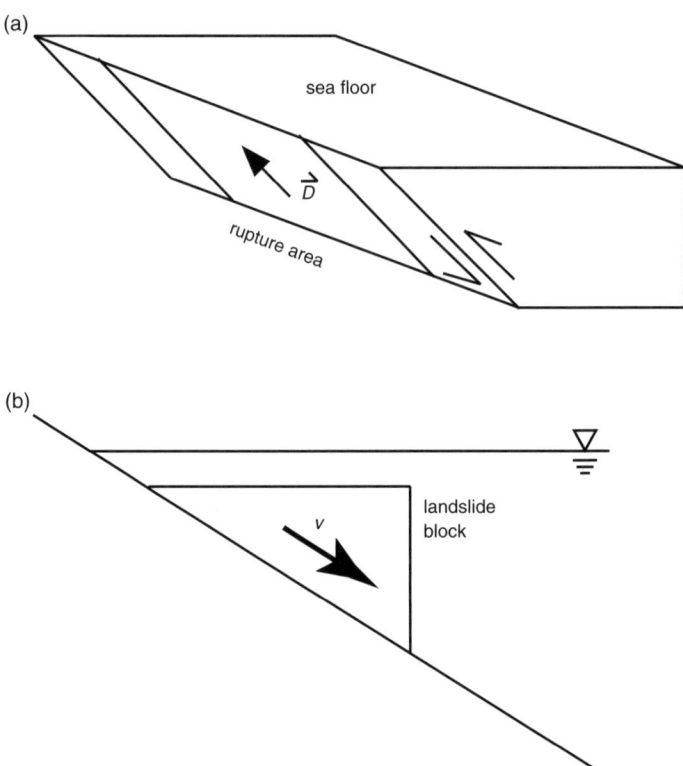

FIG. 2. Simple characterization of earthquake and landslide sources commonly used in past tsunami studies. (a) Uniform-slip vector ($\vec{D}$) within a rectangular rupture area along a planar fault. (b) Simple landslide block sliding down a constant slope with constant velocity (v).

seafloor displacement case, the leading wave profile away from the source region retains the shape of the seafloor displacement, but is reduced in amplitude by 1/2 to account for the bidirectional propagation in a 1D analysis. In contrast, for the "creeping" case, defined by Hammack (1973) as having a rate of seafloor displacement much less than the tsunami phase speed, the wave profile reflects the time history of seafloor movement rather than the overall displacement of the seafloor. To the extent that these impulsive and creeping cases can be equated to earthquake and landslide movements, spatial and temporal variations, respectively, play a dominant role in the variability of the tsunami waveform.

It is helpful to first review conventional tsunami source models used in deterministic analyses. For both seismogenic and landslide sources, the unexpected behavior of tsunamis is in reference to these deterministic models that do not accurately capture the complex details of tsunami generation physics. In the case of seismogenic tsunamis, the uniform-slip dislocation model is conventionally used, particularly in hazard assessments (Fig. 2a). This model is advantageous in that the geometric and kinematic parameters scale with seismic moment m_0 : $m_0 = \bar{\mu}\bar{D}_s\Sigma$, where $\bar{\mu}$ is the average shear modulus of rocks surrounding the rupture and $\bar{D}_s$ is the average static slip. Σ is the static rupture area (typically rectangular in shape) that is either estimated from scaling relations (e.g. see Geller (1976); Wells and Coppersmith (1994)) or from the spatial distribution of aftershocks following the earthquake. What is unaccounted for in simple tsunami generation models (and what is described below in Section 2.1) is the marked heterogeneity in the slip distribution, geometric complexities of the rupture zone, and heterogeneity in elastic moduli of rocks deforming in response to earthquake rupture.

For landslides, the conventional source model is that of a rigid block sliding down an inclined plane (Fig. 2b). This type of "wavemaker" description is convenient to implement in laboratory wave tanks for the study of various wave properties and effects on built structures. This description is also useful for benchmark comparisons of high-order numerical simulations to the wave tank measurements. Realistic landslides, however, are much more complex and diverse than the rigid block model. Various continuum (rheological and mixture theory) and granular descriptions reviewed in Section 2.2 have been used to numerically simulate landslide dynamics. Unfortunately, there are fewer data that can be used to determine the appropriate dynamic models and constitutive parameters for a given type of landslide compared with the amount of data available to develop earthquake physics models, though sample case studies are discussed in Section 3.2.

2.1. Earthquake Tsunami Generation

Tsunamigenic earthquakes typically occur at subduction zones, where two conditions related to the inter-plate thrust separating the subducting and overriding plate conspire to generate most of the world's damaging tsunamis. First, thrust motion results in higher amounts of vertical seafloor deformation in comparison to other fault mechanisms (e.g. strike slip). In particular, the dip of most inter-plate thrust faults is near the optimal dip for maximizing tsunami runup, according to the wave parameters used in runup laws (Geist, 1999). Second, the shallow dip of the inter-plate thrust fault greatly increases the possible rupture width through the seismogenic zone relative to vertical faults, for example. As a result, most of the world's largest magnitude earthquakes occur on inter-plate thrusts.

Earthquakes occurring on other faults at subduction zones have also generated devastating tsunamis in the near field (Satake and Tanioka, 1999). These include splay faults off the inter-plate thrust, thrust and reverse faults in the back arc (e.g. Sea of Japan), normal faults in the outer rise of the subducting plate caused by bending stresses (e.g. offshore Sanriku, Japan and Kuril Islands), intra-arc dip-slip faults (e.g. Mona Passage offshore Puerto Rico), and deeper earthquakes occurring within the subducting slab (Tanioka et al., 1995). Several studies have indicated connectivity among these fault types, in terms of the evolution of stress during the seismic cycle (Dmowska et al., 1988; Taylor et al., 1998). For example, it has been observed that normal faulting earthquakes in the outer rise can occur after large inter-plate thrust earthquakes, owing to stress-transfer effects (Dmowska et al., 1988; Polet and Thio, 2003). In addition, geometric complexity among these faults is often overlooked and can give rise to compound earthquake ruptures that may be difficult to classify among the fault descriptions described above (Baba et al., 2005).

In the past, it was thought that the trench-normal convergence rate was an indicator of subduction zones capable of large magnitude earthquakes (Ruff and Kanamori, 1983), and hence, large transoceanic tsunamis. However, the occurrence of the 2004 Sumatra-Andaman earthquake clearly indicates that large, tsunamigenic earthquakes can occur along inter-plate thrusts that accommodate highly oblique relative plate motion with low rates of convergence. Because the rate of trench-normal convergence along oblique subduction zones is very low, the mean return times of large earthquakes in these zones are large, resulting in temporal under-sampling in the historical record (McCaffrey, 1994; Kreemer et al., 2002; Bird and Kagan, 2004; Geist et al., 2009b).

Vertical displacement of the seafloor associated with earthquakes is directly related to slip dynamics along the fault plane, starting from a rupture initiation point (i.e. the hypocenter) and expanding over a finite area of the fault until rupture is arrested. Earthquake rupture propagation is described by the constitutive relationship that relates shear stress (τ) to slip rate ($\dot{D}$). The evolution of slip is governed by a friction law that is dependent on the effective normal stress (σ_e), that includes pore pressure effects, and a state variable (θ), commonly known as the rate- and state-friction law (Dieterich, 1979; Rice, 1983; Ruina, 1983):

$$\tau = \sigma_e \left[\mu_0 + A \ln\left(\frac{\dot{D}}{D^*}\right) + B \ln\left(\frac{\theta}{\theta^*}\right) \right], \tag{1}$$

where μ_0 is the nominal coefficient of friction and A and B are additional constitutive parameters. Earthquake rupture occurs when friction is weakening with respect to slip rate, such that $B > A$. This defines a regime of unstable sliding, in contrast to stable, aseismic sliding. The state variable represents the time evolution of stationary contacts (Dieterich, 2007) according to

$$\frac{d\theta}{dt} = 1 - \frac{\theta \dot{D}}{d_c} \tag{2}$$

(Beeler et al., 1994), where d_c is the critical slip-weakening distance (of μm scale) (see also Linker and Dieterich (1992)). The physics describing earthquake rupture therefore span many orders of magnitude in space (10^{-6} m for rupture nucleation to 10^9 m for the largest rupture length) as well as time (10^{-3} s for the breakdown leading to fault slip to 10^{10} s for the interseismic period).

The frictional stability conditions associated with the initiation of tsunami earthquakes at shallow depth are difficult to reconcile with the conventional understanding described above. A subset of tsunami earthquakes termed "slow tsunami earthquakes" by Polet and Kanamori (2000) nucleate at shallow depths along the inter-plate thrust (or subsidiary faults) and rupture up-dip to the oceanic trench. "Slow" refers to the rupture velocity and is a function of the shear modulus, which decreases substantially with depth in subduction zones (Bilek and Lay, 1999, 2000). (Other tsunami earthquakes may be attributable to the effect that triggered submarine mass movements have on the total tsunami wavefield (Kanamori and Kikuchi, 1993).) The outstanding issue associated with slow tsunami earthquakes is how very shallow rupture can occur where velocity strengthening frictional conditions ($B < A$, Eq. (1)) should exist. Tanioka et al. (1997)

and Bilek and Lay (2002) suggest that the occurrence of slow tsunami earthquakes is related to the rough topography of the downgoing plate, which may provide nucleation points for tsunami earthquakes. Seno (2002) also suggests that the conditions necessary for slow tsunami earthquakes to occur may be a transient phenomena related to the evolution of fluid pressure in or around the fault zone (cf. Taylor (1998)).

There has been a concerted effort over the past decade or so to simulate earthquake slip dynamics according to the rate- and state-friction law, although most of the work to date has focused on continental fault zones rather than inter-plate thrusts. Most of the results reveal a rich complexity in earthquake slip dynamics depending on the constitutive parameters, discretization, pre-stress distribution and fault geometry. For example, Dieterich (1995) demonstrates that for increasing values of A, slip complexity increases with multiple subevents of high slip separated by zones of little to no slip. Among earthquake mechanics models, self-healing rupture pulses are frequently distinguished from a crack-like mode of rupture (Ben-Zion and Rice, 1997; Zheng and Rice, 1998; Nielsen and Carlson, 2000). For more details regarding earthquake friction laws, fault rheology, earthquake dynamics, and fault slip, the reader is referred to topical reviews by Scholz (1998), Dieterich (2007), Rice and Cocco (2007), and Fukuyama (2009).

Elastic deformation of the rocks surrounding the fault occurs in response to the rupture process. The dynamic displacements at the Earth's surface are described by elastic wave theory (e.g. see Aki and Richards (1980)). Because shear stress vanishes at the solid-fluid interface, it is the vertical component of dynamic displacements at the seafloor that is transmitted through the water column. These displacements primarily occur as oceanic Rayleigh waves that propagate along the Earth's surface. Whereas the restoring force of tsunamis is gravity, the restoring force for Rayleigh waves is elasticity (Okal, 1988; Dahlen and Tromp, 1998; Novikova et al., 2002). Ohmachi et al. (2001) have simulated oceanic Rayleigh waves and tsunamis using a coupled model of dynamic displacements and showed that the maximum water heights are significantly greater above the source region, but approximately the same in the far field, in comparison to static displacements. Because oceanic Rayleigh waves are coupled with the overlying water column of the ocean, they propagate at phase and group speeds for elastic Rayleigh waves in the solid earth. Furthermore, because the solid earth and ocean are coupled during Rayleigh wave propagation, no runup occurs for these waves. In contrast, tsunamis are "free" (i.e. mostly uncoupled) gravity waves in the water column that propagate at phase and group speeds according to the long-wave hydrodynamic theory. Directly above the source region, however, wave heights may be significantly augmented by dynamic displacements.

Thus, tsunami generation relates more to the "static" displacement of the seafloor in response to earthquake rupture in contrast to seismic ground motion that depends on the time evolution of rupture. Static in this sense means final displacement in a reference frame prior to rupture nucleation. This can be also interpreted as "permanent" deformation (as in the formation of fault scarps, for example) on the time scale of tsunami propagation (but not necessarily on the time scale of the earthquake cycle). In seismology, the static displacement field can be derived from standard formulas for dynamic displacements with $t \to \infty$ (pg. 84, Aki and Richards (1980)).

Static elastic displacement at a point on the seafloor $\mathbf{r}(x, y)$ is the result of elastic deformation in response to coseismic slip $D(\xi, y)$ integrated across the entire fault plane

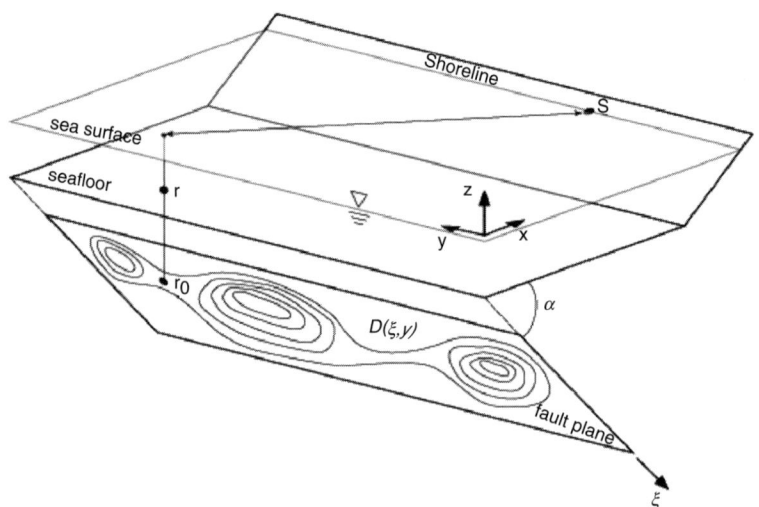

FIG. 3. Schematic perspective view of the tsunami problem. Slip on fault plane $D(\xi, y)$ results in coseismic displacement at a point on the seafloor $u_m(\mathbf{r})$. This in turn results in an initial disturbance at the sea surface that propagates to the shoreline as the tsunami observed at point **s**.

(Fig. 3). The three components of displacement ($m = 1, 2, 3$) at **r** can therefore be expressed as a function of slip at a point on the fault $\mathbf{r}_0$ (Rybicki, 1986)

$$u_m(\mathbf{r}) = \iint_\Sigma D_i(\mathbf{r}_0) v_j(\mathbf{r}_0) U_m^{ij}(\mathbf{r}, \mathbf{r}_o) d\Sigma \tag{3}$$

where $\nu(\mathbf{r}_0)$ is the surface normal of a fault plane with arbitrary geometry and Σ is the area of the fault plane. For an isotropic medium, the six Green's functions are given by

$$U_m^{ij} = U_m^{ji} = \lambda \delta_{ij} G_m^{n,n} + \mu [G_m^{i,j} + G_m^{j,i}], \tag{4}$$

where λ is Lamé's first constant and δ_{ij} is the Kronecker delta. The static elastic Green's function has known analytic solutions only for a few simple slip distributions and fault geometries (e.g. see Singh *et al.* (1994)). Because Eqs (3) and (4) are based on the static approximation, temporal changes in slip and seafloor deformation during rupture propagation and slip evolution are not accounted for. Often parameters such as rupture propagation speed and rise time are used in a simple kinematic framework as in the Haskell rupture model (Haskell, 1964), particularly for very long ruptures such as the 2004 Sumatra-Andaman earthquake (Hirata *et al.*, 2006; Fujii and Satake, 2007). Recently, dynamic rupture propagation models have been employed that provide a physically consistent framework to predict the spatial and temporal heterogeneity of slip (e.g. see Wendt *et al.* (2009)).

2.2. Landslide Tsunami Generation

Generation of tsunamis by submarine landslides is a complex process that occurs through distinct temporal phases: failure, post-failure dynamics (e.g. debris flow), and turbulent boundary layer flow (i.e. turbidity currents). Because most tsunamigenic landslides are triggered by earthquakes, one has to consider inertial displacements of the uppermost compliant layer in response to strong-ground motion. Given that the conditions are met for failure to occur, it is likely that a shear dislocation develops at the base of the landslide and propagates in all directions to form the eventual head scarp (up slope) and initial sliding plane (down slope) (Martel, 2004; Petley et al., 2005). Our primary concern for tsunami generation is the dynamics of the failed mass, which is described below. During the latter stages of mass movement failure, turbidity currents most often form. Because this process is characterized primarily by turbulent boundary layer flow and involves relatively smaller changes in seafloor displacement, it is usually not influential in the tsunami generation process.

Many and diverse types of landslides have been classified according to their physics and subaerial observational evidence. The most well known classification scheme is that of Varnes (1978). Since then, there has been a concerted effort to develop a detailed taxonomy of landslides, particularly of the flow type (Hungr et al., 2001). The basic types of mass movements include slow-moving earth flows (sometimes termed landslides or slides in a specific connotation of the terms), topples, spreads, falls, and fast-moving flows (Locat and Lee, 2002). The concern for tsunami generation is primarily fast-moving submarine debris avalanches and debris flows. A recent classification of debris flows proposed by Coussot and Meunier (1996) is based on two criteria directly related to the mechanical properties of debris flows: solid/water fraction and material type (cohesive versus granular). This classification is seen as an important foundation for understanding the complex dynamics of debris flows that are governed by non-Newtonian rheologies.

In understanding the physics of debris flows, two approaches have primarily been considered: viscoplastic fluid and mixture (or granular) theory. Viscoplastic fluid models have been used to describe submarine muddy debris flows, whereas mixture theory has been used primarily to describe subaerial granular or two-phase flows. The latter has recently been incorporated into a coupled model for tsunami generation (Fernández-Nieto et al., 2008). In adapting subaerial landslide dynamic models for the submarine environment, the main effect of having water as the ambient fluid, rather than air, is a reduction in gravitational forcing, owing to buoyancy.

Because most tsunamigenic mass movements along continental slopes involve predominantly fine-grained sediment, the viscoplastic fluid model for muddy debris flows is a central model for tsunami generation. A clay content of only 10% or more is needed for debris flows to be adequately modeled by viscoplastic rheology (Coussot and Meunier, 1996; Coussot et al., 1998). The longitudinal momentum equation for laminar flow is given by

$$\rho_m \left(\frac{\partial u_x}{\partial t} + u_x \frac{\partial u_x}{\partial x} + u_z \frac{\partial u_x}{\partial z} \right) = \rho_m g \sin\theta - \frac{\partial p}{\partial x} + \frac{\partial \tau}{\partial z}, \qquad (5)$$

where u_x and u_z are the velocity components in the x (down slope) and z directions respectively, ρ_m is the density of the mud flow, $p(x, z, t)$ is the pressure (assumed to be hydrostatic), θ is the slope angle, and $\tau(x, z, t)$ is the shear stress in the mud flow (Jiang and Leblond, 1993; Imran et al., 2001). The continuity equation for this system is

$$\frac{\partial u_x}{\partial x} + \frac{\partial u_z}{\partial z} = 0. \tag{6}$$

Various nonlinear constitutive relations have been used to relate shear strain rate to shear stress for muddy debris flows (Coussot, 1997). These include Bingham plastic, Herschel-Bulkley, and bilinear rheologies. The Bingham plastic fluid is characterized by a finite yield stress (τ_y) such that,

$$\mu\gamma = \begin{cases} 0 & \text{if } |\tau| < \tau_y \\ \tau - \tau_y \mathrm{sgn}(\gamma) & \text{if } |\tau| \geq \tau_y, \end{cases} \tag{7}$$

where μ is dynamic viscosity and γ is the strain rate ($\gamma = \frac{\partial u}{\partial z}$). The Herschel-Bulkley rheology is a power-law rheology (i.e. non-Newtonian):

$$K\gamma^n = \begin{cases} 0 & \text{if } |\tau| < \tau_y \\ \tau - \tau_y \mathrm{sgn}(\gamma) & \text{if } |\tau| \geq \tau_y, \end{cases} \tag{8}$$

where n and K are material-specific constants. When $n = 1$, the Herschel-Bulkley rheology is equivalent to the Bingham plastic rheology. The bilinear rheology (Locat, 1997) involves two viscous regimes of flow described by dynamic viscosities at low and high strain rates (μ_l and μ_h, respectively where $\mu_h < \mu_l$):

$$\tau = \left[\tau_{ya} + \mu_h |\gamma| - \frac{\tau_{ya}\gamma_0}{|\gamma| + \gamma_0}\right] \mathrm{sgn}(\gamma), \tag{9}$$

where τ_{ya} is the apparent yield stress relative to the high strain rate regime and γ_0 is the reference strain rate given by

$$\gamma_0 = \frac{\tau_{ya}}{\mu_l - \mu_h}. \tag{10}$$

For these nonlinear rheologies, the no-slip boundary condition along the base of the debris flow results in two flow zones: a shear zone at the base of the flow where the shear stress is greater than the yield stress and a plug zone above where the yield stress is not exceeded. The boundary between the two zones is termed the yield interface. In formulating a solution to the momentum equations, the horizontal and vertical velocities and the horizontal velocity gradient are constrained to be continuous across the yield interface (Jiang and Leblond, 1993).

To model granular mass movements that have a smaller proportion of fine sediment and water, mixture theory that was developed for subaerial debris avalanches (Iverson and Denlinger, 2001) has been adapted in a few cases for the submarine environment (Fernández-Nieto et al., 2008). For a two-phase solid/fluid mixture in which the fluid velocities and accelerations differ negligibly from those of the solids, the momentum equation is

$$\rho_m \left(\frac{\partial \mathbf{v_s}}{\partial t} + \mathbf{v_s} \bullet \nabla \mathbf{v_s} \right) = \rho_m \mathbf{g} - \nabla \bullet (\mathbf{T_s} + \mathbf{T_f}), \tag{11}$$

where $\mathbf{T}_s$ and $\mathbf{T}_f$ are the stress tensors for the solid and fluid phases, respectively, and the density of the mass movement is calculated from the volume fractions (V_s and V_f) of each phase $\rho_m = \rho_s V_s + \rho_f V_f$. The constitutive theory used is intergranular Coulomb friction, modified by pore pressure and Newtonian viscous fluid stresses. For high enough pore pressures, internal friction is greatly reduced and the mass behaves viscously. It has been argued that mixture theory has limited application in the submarine environment because pore pressure diffusion is likely to be minimal and because a finite yield strength is needed to explain the thickness and mid-slope termination of many submarine debris avalanches (Coussot and Meunier, 1996; Elverhøi *et al.*, 2005). However, mixture theory may be applicable in specific geologic environments such as carbonate and volcanic-dominated islands, away from continental sources of clay.

2.3. Tsunami Propagation

Large-scale changes in ocean surface elevation induced by earthquakes, landslides, and other geologic phenomena present quasi-initial conditions for tsunami propagation. The relevant non-dimensional wave parameters are

$$\varepsilon = \frac{a_0}{h_0}, \quad \mu = \frac{h_0}{\ell_0}, \quad \delta = \frac{\Delta h}{h_0}, \tag{12}$$

where a_0 is the characteristic wave amplitude, h_0 the characteristic water depth, ℓ_0 the characteristic wavelength, and Δh is the change in seafloor depth. When the dispersion parameter μ is small ($\ll 1$), the shallow-water wave equations are appropriate to describe tsunami propagation. These equations include the continuity equation

$$\frac{\partial(\eta + h)}{\partial t} + \nabla \bullet [\mathbf{v}(\eta + h)] = 0 \tag{13}$$

and the momentum equation

$$\frac{\partial \mathbf{v}}{\partial t} + (\mathbf{v} \bullet \nabla)\mathbf{v} + g\nabla \eta = 0, \tag{14}$$

where h is the water depth (a function of time owing to source dynamics), η is the wave elevation, and $\mathbf{v}$ is the depth-averaged horizontal vector velocity field. When dispersion cannot be neglected, wave components are separated such that different frequencies propagate at different speeds (i.e. frequency-dependent phase speed). For small ε, the nonlinear convective inertia term $(\mathbf{v} \bullet \nabla)\mathbf{v}$ in Eq. (14) above can be ignored, resulting in the linear long-wave equations. Conventionally, the linear form is valid in deep water until the water approaches the shore (i.e. $\varepsilon \to 1$). However, as noted by Lynett and Liu (2002), nonlinear effects are present in the source region for landslide tsunami generation where seafloor displacement can be of the same order as the water depth. The reader is referred to other publications that comprehensively describe the physics of tsunami waves (Carrier, 1971; Mei, 1989; Madsen and Schäffer, 1998; Liu, 2008; Lynett, 2008).

For earthquakes, the initial tsunami amplitudes in the source region are typically less than those for tsunamigenic landslides. However, landslide-generated tsunamis have much smaller source dimensions and are more affected by frequency dispersion, leading to a greater attenuation of the initial wave amplitude in the far field. Therefore, whereas seismogenic tsunamis can have destructive influences at far-field distances, the impact of landslide-generated tsunamis is limited to the near field.

A particular aspect of tsunamis that is critical in analyzing the data that follow is the coastal response of tsunamis, including the excitation of edge waves and runup. Edge waves are a particular type of coastal wave trapped by refraction that propagate parallel to the coastline. In combination with scattering and resonance resulting from propagation along an irregular coastline, edge waves create a complex waveform in which the offshore amplitude, runup, and timing of the largest wave are difficult to predict.

The theoretical understanding of edge waves is based on simple shelf and slope geometries. Edge waves occur in distinct modes (n), with the fundamental mode ($n = 0$), also known as the Stokes mode, which is the most commonly observed. For a semi-infinite sloping beach of slope β, the dispersion relation is (Ursell, 1952; Liu et al., 1998)

$$\omega_n^2 = gk_n \sin(2n+1)\beta, \quad (2n+1)\beta < \pi/2. \tag{15}$$

Snodgrass et al. (1962) developed the propagation characteristics of edge waves along a flat shelf and distinguished between the discrete edge wave modes and leaky modes that occupy a continuous spectrum. Ishii and Abe (1980) considered a more complex case of edge waves along a stepped continental margin profile with an intervening linear slope and compared their results (dispersion relation and amplitude of the fundamental-mode edge wave) with those from the vertical step profile. The amplitude of edge wave modes for a semi-infinite sloping beach is based on Laguerre polynomials of order n ($L_n(x)$) and is of the form (González et al., 1995)

$$\eta_n(x, y) = Ae^{i(k_n y - \omega_n t)} e^{-k_n x} L_n(x). \tag{16}$$

The cross-shore wave profiles for the first few edge modes are shown in Fig. 4. For the more general case, the amplitude function is a solution to the confluent hypergeometric equation (Kummer's equation) (Ishii and Abe, 1980; Mei, 1989).

Runup describes the evolution of tsunami waves as they progress up a beach slope. Analytically, it is a particularly difficult hydrodynamic problem to solve, owing to nonlinearity near the runup front (Liu et al., 1991) and moving boundary conditions (Kennedy et al., 2000; Lynett et al., 2002). Carrier and Greenspan (1958) provided a general solution to the problem, using a hodograph coordinate transformation that results in a linear form of the wave equation. A later study (Carrier, 1995) looked specifically at tsunami runup of both the broadside and edge waves from a near-field source. Analytic runup laws have been derived for specific waveforms, such as solitary

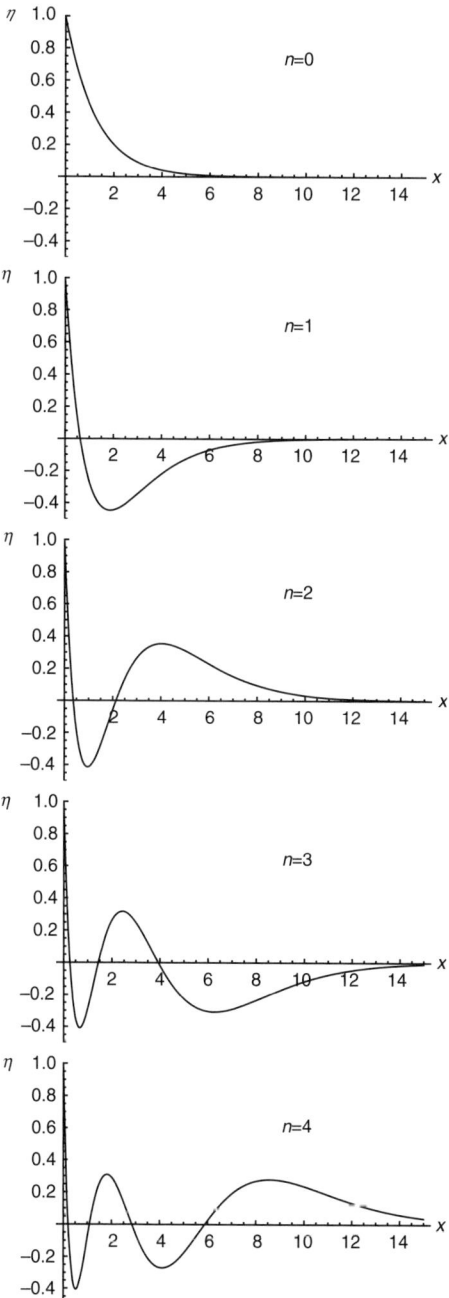

FIG. 4. Cross-shore wave profiles of the first five edge wave modes for a semi-infinite sloping beach (Eq. (16)).

waves (Synolakis, 1987) and *N* waves (Tadepalli and Synolakis, 1994). Recently, Madsen *et al.* (2008) has examined solitary wave theory in detail as it applies to tsunamis. Recent studies have also generalized the Carrier and Greenspan transform to include arbitrary incident waveforms (Carrier *et al.*, 2003; Tinti and Tonini, 2005). Using runup theory to explain field observations is complicated by variable offshore topographies (Kanoglu and Synolakis, 1998). Specific methods have also been developed to handle turbulence that may be highly significant for large tsunamis (most notable observed for the 2004 Indian Ocean tsunami), both in the form of bottom boundary layer flow (parameterized by bottom friction) and for cases where the tsunami breaks offshore and develops into a bore (Yeh, 1991; Chen *et al.*, 2000; Kennedy *et al.*, 2000; Lynett *et al.*, 2002). A recent review of tsunami runup and inundation is given by Yeh (2008) and Lynett (2008).

3. Near-field Regimes: Broadside and Oblique

In this section, I examine the phenomenology of tsunami events in the near field. First, source observations that directly relate to tsunami generation are reviewed. The focus here will be on the spatial distribution of coseismic slip along the fault plane for seismogenic tsunamis, deduced from inversions of seismic, geodetic, and tsunami data that have a primary effect at near-field distances. I then examine how source variations affect the near-field tsunami, both in coastal regimes broadside from and oblique to the rupture zone. In this respect, the approach can be considered as a stochastic initial boundary value problem as defined by Leblond and Mysak (1978) in which the spatial distribution of the wave forcing function is random, as are the propagation characteristics along the coast. In the broadside regime, I examine how spatial distributions in runup are causally related to spatial distributions of slip on the fault. In the near-field oblique regime, I examine the importance that edge wave propagation has in explaining tsunami observations.

3.1. Source Observations—Earthquakes

Data used to determine kinematic rupture processes for tsunamigenic, subduction zone earthquakes primarily include waveforms recorded by strong motion and broadband seismic instruments. Inverse methods can be performed on this data to determine the slip history during rupture. Seismic inversion methods include inverting body waves (Kikuchi and Kanamori, 1991), surface waves (Ihmlé, 1996a) and wavelet methods to simultaneously invert seismic and geodetic data (Ji *et al.*, 2002). The inversion of seismic data requires assumptions of fault geometry and other fixed parameters such as crustal structure (Beresnev, 2003). The fault orientation is often available from seismicity studies, centroid moment tensor inversion (e.g. see Tsai *et al.* (2005)) or from focal mechanism determination (e.g. first motions), although for the latter, it is difficult to constrain the dip of shallow-angle faults, such as inter-plate thrusts, owing to the steep take-off angle of teleseismic *P* waves (Michael and Geller, 1984). Rupture length and width are often estimated from the aftershock distribution. Other difficulties arise because of the heterogeneous velocity structure associated with subduction zones (Engdahl *et al.*, 1982; Kisslinger, 1993). Despite these limitations, there is a great deal of information available from seismic inversions on the slip distribution of past tsunamigenic earthquakes.

Geodetic measurements and tsunami waveforms provide additional data to resolve the static slip distribution of earthquakes. The availability of high-precision GPS data greatly increases the amount of data one can use to determine slip distribution, although earlier leveling data is valuable for historic earthquakes as well. Care must be taken in processing geodetic data to account for after-slip and post-seismic earth responses (Heki and Tamura, 1997; Miyazaki et al., 2004; Hashimoto et al., 2006; Pollitz et al., 2008). The direct arrival of tsunami waveforms also provides valuable data to invert for slip distribution (e.g. see Fujii and Satake (2007); Piatanesi and Lorito (2007)). As will be discussed in Section 4.1, the direct unobstructed part of the tsunami waveform that is analyzed in tsunami inversions is quite stable during propagation and not subject to random processes that, for example, dictate maximum tsunami amplitude that can occur later in the tsunami coda.

Seismic and geodetic inversions resolve slip best toward shore, where the observation stations are located, whereas tsunami inversions resolve slip best in deeper water (i.e. at shallow depths below the seafloor for inter-plate thrust earthquakes) (Sagiya and Thatcher, 1999). Therefore joint inversions of the aforementioned data sets are particularly useful to fully resolve slip throughout the rupture zone of large earthquakes (e.g. see Johnson et al. (1996); Thio et al. (2004); Ichinose et al. (2007); Rhie et al. (2007)). However, to avoid any apparent circular reasoning in this paper involving slip that could be determined from the tsunami data I am describing, I focus primarily on slip distributions determined from seismic waveform inversions.

Observed slip distributions are spatially complex and are considered self-affine, to the extent provided by the resolution of the inverted data. This characteristic of coseismic slip is a result of rupture dynamics, either through fractal pre-stress conditions, or from fault roughness (cf. Andrews (1980); Power and Tullis (1991); Oglesby and Day (2002)). Several different statistical models have been developed to describe these observations. The slip distribution model initially developed by Andrews (1980) and further discussed by Herrero and Bernard (1994) is characterized by a power-law decay in the wavenumber spectrum beyond a corner wavenumber k_c:

$$D(k) = C \frac{\Delta\sigma}{\mu} \frac{L}{k^2} k > k_c \tag{17}$$

where $\Delta\sigma$ is the mean stress drop, μ the shear modulus, L the rupture dimension and C a constant. It has been shown that the wavenumber decay exponent can be linked to the spectral decay of far-field seismic displacement amplitudes in the frequency domain (Hanks, 1979; Frankel, 1991; Hisada, 2000, 2001). Therefore there are two different types of observations that can be used to constrain the stochastic slip model: seismic displacement spectrum and seismic inversions. For the former, there have been several studies that define the spectrum of inter-plate thrust earthquakes (e.g. see Hartzell and Heaton (1985)). Interestingly, given the unique behavior of slow tsunami earthquakes in many other respects, Polet and Kanamori (2000) find similar spectral decay exponents as for typical inter-plate thrust earthquakes that rupture at deeper focal depths.

Seismic inversion results have been used to determine spectral decay of slip for the more general case of $D(k) \propto 1/k^{\nu+1}$ for a 2D fault, where ν controls the long-range correlation or "roughness" of the slip distribution. Tsai (1997a) estimated optimal values of ν for six different onshore earthquakes (most of them with two different published inversion solutions using different methodologies). Mai and Beroza (2002) refined the statistical estimation by considering slip distribution from the standpoint of several different spectral decay laws, including the fractal model (Eq. (20)), Gaussian, exponential, and von Kármán autocorrelation functions. They found that the von Kármán autocorrelation function provides the best fit, though slip distributions are also well described by the fractal distribution. The von Kármán autocorrelation function used by Mai and Beroza (2002) includes different correlation lengths in the strike and the dip direction (a_x and a_z, respectively), such that

$$D(k) \propto \frac{a_x a_z}{\left(1+k^2\right)^{\nu+1}}. \tag{18}$$

(The autocorrelation function is related to the power spectrum via the Fourier transform according to the Wiener-Khintchine theorem, e.g. see Andrews (1980).) In comparing the slip distribution models to seismic inversion results, Mai and Beroza (2002) found a scale-invariant value for the spectral decay exponent that is greater than 2.

Lavallée et al. (2006) considered the possibility that slip fluctuates more than the standard stochastic model described above. The stochastic slip distribution model is fully defined by the wavenumber spectrum ($D(k)$ above) and a random function $R(x)$, such that

$$D(x) = D_0 F^{-1}\left[R_s(k) D(k)\right], \tag{19}$$

where $F^{-1}[\bullet]$ is the inverse Fourier transform, $R_s(k)$ is the Fourier transform of $R(x)$ and D_0 is a constant (Liu-Zeng et al., 2005). Typically, $R(x)$ is specified by a Gaussian distribution with zero mean. As indicated by Lavallée et al. (2006), $D(k)$ specifies the long-range correlation of the slip function whereas $R(x)$ specifies its variability.

Lavallée and Archuleta (2003) first noted the variability of slip associated with the 1979 Imperial Valley earthquake from inversion of near-field strong motion data (Archuleta, 1984) and indicated that a distribution family more general than the Gaussian should be considered to encompass this high variability. The Lévy α-stable distribution is considered in which the characteristic function for the probability distribution is given by (Samorodnitsky and Taqqu, 1994)

$$j(k) = \begin{cases} \exp\left\{ik\mu - c^\alpha |k|^\alpha \left[1 - i\beta \mathrm{sgn}(k) \tan\frac{\pi\alpha}{2}\right]\right\} & \text{if } \alpha \neq 1 \\ \exp\left\{ik\mu - c|k|\left[1 - i\beta\frac{2}{\pi}\mathrm{sgn}(k) \ln|k|\right]\right\} & \text{if } \alpha = 1 \end{cases}, \tag{20}$$

where α is a stability parameter ($0 < \alpha \leq 2$), μ a shift parameter (real number), β a skewness parameter ($-1 \leq \beta \leq 1$), and c a scale parameter ($c \geq 0$). For $\alpha = 2$, this becomes the characteristic function for the Gaussian distribution with mean

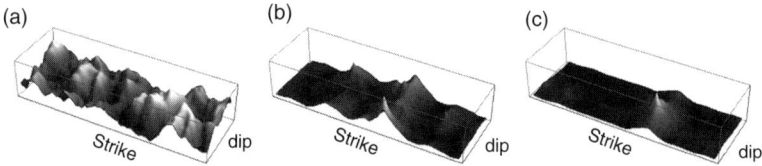

FIG. 5. Example of stochastic slip distributions using three different distributions for $R(x)$: (a) Gaussian; (b) Cauchy; (c) Lévy. The power spectrum $D(k)$ is the same in each case.

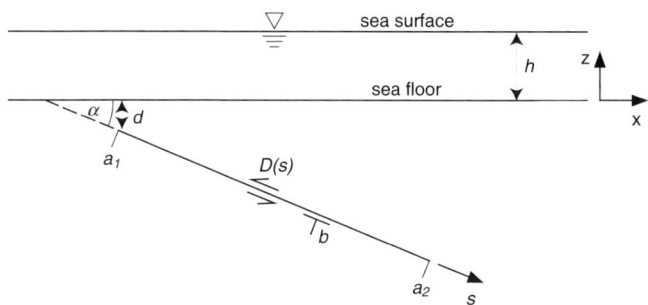

FIG. 6. Schematic and coordinate system for calculating elastic displacement from an edge dislocation on an inclined fault.

μ and variance $2c^2$. Examples of distributions with heavier density tails include the Cauchy distribution ($\alpha = 1$ and $\beta = 0$) and the Lévy distribution ($\alpha = 1/2$ and $\beta = 1$). Figure 5 compares different slip distributions using the same wavenumber spectrum but different distributions for $R(x)$ (Gaussian, Cauchy, and Lévy). The heavy-tail distributions (Fig. 5b, c) produce more extreme values of slip, relative to the average slip within the rupture zone. Lavellée et al. (2006) estimated distribution parameters (α, μ, β, and c) and the wavenumber decay exponent ($\nu + 1$) for several earthquakes, including the predominantly dip-slip 1994 Northridge event. Most earthquakes analyzed have α-values less than 2, and therefore have heavy tails.

Although there are very few direct observations of static displacement of the seafloor, displacement can be computed from the dislocation theory if a statistical description of coseismic slip is known, such as that described above. The general form of the coseismic displacement equation (Eqs (3) and (4)) is applied for cases that result in tsunami generation. For tsunami applications, the focus is on the vertical component of displacement, although horizontal displacement can be calculated in a similar fashion. In addition, inclined dip-slip faults such as the inter-plate thrust of subduction zones are of primary interest. Surface displacement for an edge dislocation on an inclined fault with Burger's vector b as shown in Fig. 6 has been derived by several authors (e.g. see Dmowska and Kostrov (1973); Freund and Barnett (1976); Rudnicki and Wu (1995)). Freund and Barnett's (1976) expression using the geometric parameters shown in

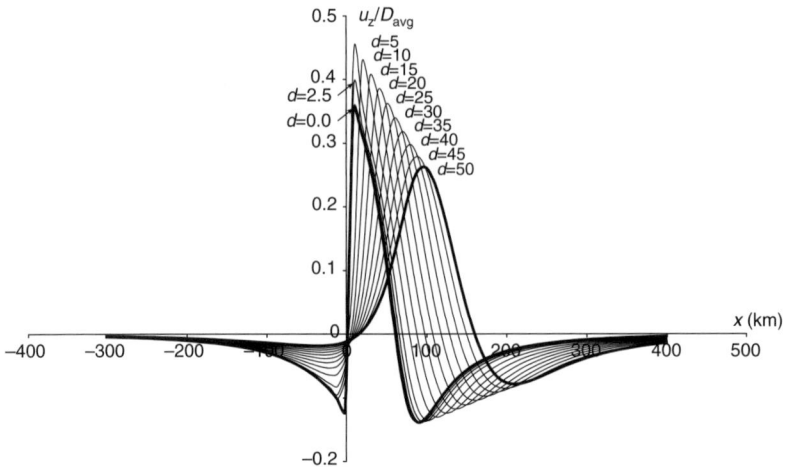

FIG. 7. Vertical displacement profiles from a crack along the inclined fault plane specified in Fig. 6 as a function of depth d to the top of rupture (in km). Bold lines represent end-member cases ($d = 0$ km and 50 km). Vertical uplift u_z is normalized with respect to average slip D_{avg}. Rupture width = 80 km, length = 200 km, and dip (α) = 30°.

Fig. 6 is

$$u_z(x, s) = b U_z(x, s), \tag{21}$$

where

$$U_z(x, s) = \frac{xs \sin^2 x}{\pi(x^2 + s^2 - 2xs \cos\alpha)} + \frac{\sin\alpha}{\pi} \tan^{-1}\left(\frac{x - s\cos\alpha}{s\sin\alpha}\right). \tag{22}$$

For a crack of finite width and distributed slip (i.e. dislocation density), expressions for displacement at the seafloor are obtained by integrating Burger's vector solutions (see Dmowska and Rice (1986) for a summary). The integral form for vertical displacement given by Freund and Barnett (1976) is

$$u_z(x, 0) = \int_{a_1}^{a_2} U_z(x, s) D'(s) ds. \tag{23}$$

Variation in vertical displacement profiles as a function of source depth is shown in Fig. 7. These results show that for a given slip distribution, the vertical displacement is characterized by a two-sided displacement field in the strike-normal (wave profile) direction: uplift in the up-dip direction and subsidence in the down-dip direction. This transformation from one-sided slip (i.e. positivity constraint) to two-sided vertical displacement is linked to the slip gradient term $D'(s)$ in Eq. (23) above. As the depth to slip on the fault becomes shallower (i.e. toward the seafloor), the vertical displacement profile becomes sharper. When slip on the fault breaks through to the seafloor, a scarp is formed (that is correspondingly attenuated in the water column). The conditions for which this would occur are discussed by Rudnicki and Wu (1995). In general though, the static elastic Green's function in essence acts as a low-pass spatial filter of the slip distribution in calculating seafloor displacements (cf. Figure 11 in Geist and Dmowska (1999)). Green's function also effectively transforms the one-sided slip functions to a two-sided displacement profile in the dip direction (uplift and subsidence).

Within the water column, the tsunami Green's function derived by Kajiura (1963) relates seafloor displacement to ocean surface elevation. In effect, short wavelengths of the seafloor displacement field are filtered through the water column by a factor of $1/\cosh(kh)$. For most earthquakes, this does not result in a significant attenuation from seafloor displacement to generation of the initial tsunami wavefield. However, fault scarps and anomalously short-wavelength components of the displacement field will be filtered out in deep water during tsunami generation. 3D simulations of tsunami generation by Saito and Furumura (2009b,c) indicate that horizontal deformation wavelengths of the seafloor must be at least 10 times the water depth for the surface height to mimic vertical seafloor displacement (within approximately 10%).

For inter-plate thrust earthquakes, it is important to recognize the strong structural heterogeneity in rock types and elastic properties. Sediment that is accumulated on the downgoing plate is accreted onto and underplated beneath the overriding plate. Typically the inter-plate thrust of a subduction separates high modulus igneous rocks below from low modulus sedimentary rocks above, though the fault may not be exactly co-located with the contact between rock types. Ma and Kusznir (1994) and Savage (1998) investigated the effect that horizontal elastic layering has on surface displacements through the derivation of analytic solutions to the governing equations. For more complex cases, finite-element models have also been widely used to account for elastic heterogeneity in the calculation of surface displacement fields (e.g. see Yoshioka et al. (1989); Cattin et al. (1999); Masterlark et al. (2001); Masterlark (2003); Zhao et al. (2004); Sato et al. (2007)). For example, the effect of elastic heterogeneity has been investigated for potential tsunamigenic earthquakes ($M_w = 6.5$) on the Seattle fault (Fig. 8a, b), which trends up-dip from high rigidity basement rocks into low rigidity sedimentary rocks of the Seattle basin (Geist and Yoshioka, 2004). Figure 8c shows a comparison of the vertical displacement field using an elastic half-space and using elastic constants specific to basement and sedimentary basin rocks.

There are very few measurements of seafloor deformation arising from earthquake rupture for testing the aforementioned models. One exceptional record, however, is the measurement of vertical displacement by a bottom pressure recorder that is part of a cabled deep-sea observatory (Hirata et al., 2002) during the 2003 Tokachi-Oki earthquake (Baba et al., 2006b; Mikada et al., 2006; Nosov and Kolesov, 2007). Rapid vertical displacement of the seafloor excited low-frequency elastic oscillations in the water column that dominated the pressure observations (Nosov and Kolesov, 2007). Measured coseismic displacement, after filtering out the water-column oscillations, was consistent with the elastic displacement models of the event (Mikada et al., 2006), using a slip distribution inverted from the observed seismograms of the earthquake (Yamanaka and Kikuchi, 2003; Yagi, 2004). Other than this notable example, observations of both coseismic slip (from seismic inversions) and vertical displacement at the Earth's surface for a single event are only available for onshore earthquakes and not for submarine events.

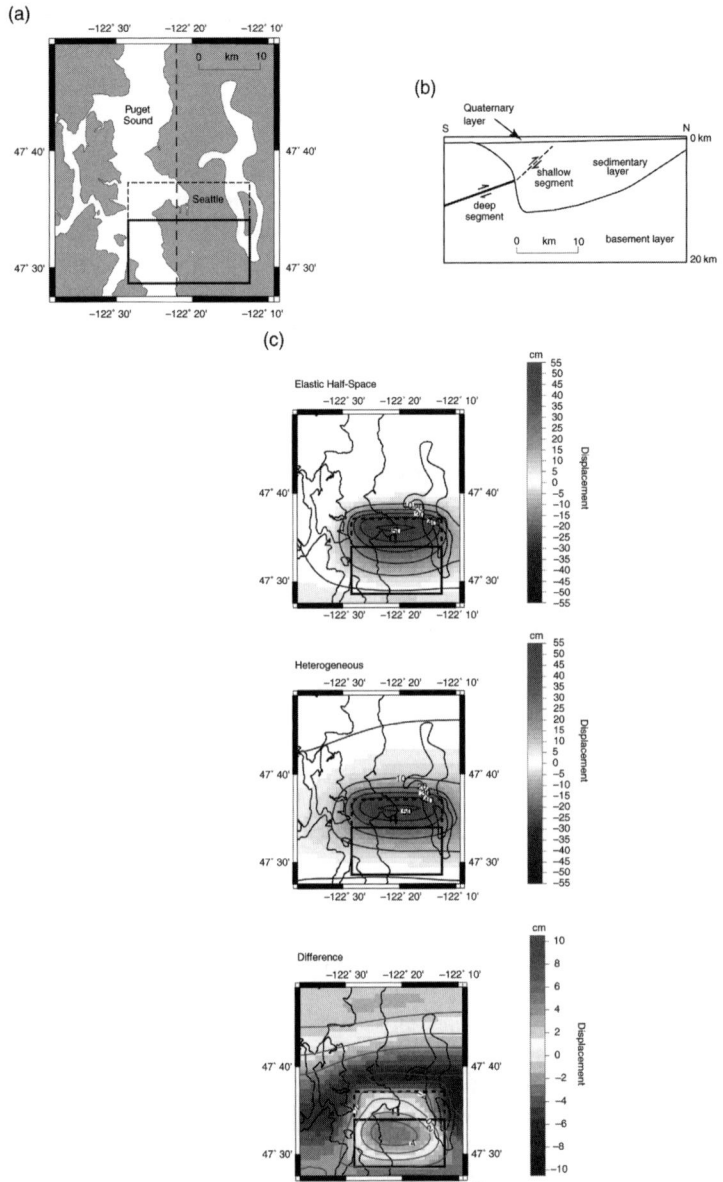

FIG. 8. (a) Mapview of Puget Sound with projection of deep (bold rectangle) and shallow (dashed rectangle) segments of the Seattle fault. (b) Cross section showing the structure of Seattle sedimentary basin. Slip on the Seattle fault extends from basement rocks up-dip into the Seattle basin. (c) Comparison of vertical displacement fields from a homogeneous elastic structure (half-space) and elastic heterogeneity arising from the structure of the Seattle basin (Geist and Yoshioka, 2004). Difference between the two displacement fields shown at the bottom.

The lack of seafloor displacement measurements precludes a direct determination of whether rupture can occur along splay faults that are synthetic to the inter-plate thrust. The possibility of splay fault rupture has been proposed to explain a greater than expected near-field tsunami for a given amount of slip (e.g. see Fukao (1979); Cummins and Kaneda (2000)). Splay faults in the hanging wall of inter-plate thrusts have been imaged by seismic-reflection profiles for specific subduction zones, typically in the accretionary wedge (e.g. see Park et al. (2002); Moore et al. (2007); Sibuet et al. (2007)). Because seismic waveforms cannot routinely detect branching rupture paths for inter-plate thrust earthquakes, most of our knowledge as to whether splay fault rupture can occur comes from an understanding of fault mechanics. Several recent studies indicate that splay fault rupture can only occur under specific pre-stress conditions and branch angles (Poliakov et al., 2002; Kame et al., 2003; Bhat et al., 2007; Templeton et al., 2009). In addition, barriers to inter-plate rupture in the strike direction may create the necessary conditions for rupture to jump to a splay fault segment (Wendt et al., 2009). Our primary observational evidence of splay fault rupture is from the inversion of near-field tsunami waveforms (Baba et al., 2006a).

3.2. Source Observations—Landslides

Unfortunately, there is no routine instrumental monitoring of submarine landslides analogous to seismograms such that one could directly infer the dynamics and kinematics of landslide movement. There have been some studies of seismic waves generated by energetic landslides (both submarine and subaerial) that are worth noting. Landslides can be distinguished from earthquakes on seismograms by a depletion of high-frequency energy (e.g. see La Rocca et al. (2004)) and are represented by a horizontal point force at the surface (Dahlen, 1993), rather than a double-couple system of forces used to represent an earthquake mechanism. Recently, an unusual seismic recording of an event in the Gulf of Mexico may be ascribed to a submarine landslide according to these characteristics (Nettles, 2007). Furthermore, Brodsky et al. (2003) demonstrated that the basal friction of landslides can be estimated from a given source-time function (Kanamori et al., 1984), concluding that basal friction does not appear to significantly vary for the three subaerial volcanogenic landslides they analyzed.

To some extent, tide gage records can be used to constrain the location and constitutive parameters of the landslide (as described in Section 2.2), although there are very few confirmed instrumental recordings of landslide tsunamis. One example is the 1929 Grand Banks tsunami that possibly had a landslide-generation component. The triggering earthquake and geomorphology of the submarine failure have been described in several studies (e.g. see Hughes Clarke (1990); Bent (1995); Piper et al. (1999)). From these descriptions and using simplified assumptions of uniform thickness and Newtonian viscosity, Fine et al. (2005) were able to show that the tsunami resulting from the landslide is consistent with travel time observations for the initial wave and the amplitude as recorded at the Halifax, Nova Scotia tide gage station (see also Trifunac et al. (2002)). Another example is the landslide in Monterey Canyon triggered by the onshore 1989 Loma Prieta earthquake that generated a tsunami recorded by the nearby Monterey tide gage station (Ma et al., 1991). The timing of the direct arrival and the waveform characteristics as observed on the tide gage record are used to determine the approximate location of the landslide (near the canyon head) and a low yield stress using the bilinear constitutive relation (Eqs (9) and (10)), respectively (Geist *et al.*, 2009a). Both the 1929 Grand Banks and 1989 Monterey Canyon debris flows were accompanied by vigorous turbidity currents in their aftermath.

3.3. Tsunami Observations—Near-field Broadside Regime

I turn now to the effect that spatially heterogeneous slip and displacement of the seafloor that was discussed in Section 3.1 has on near-field tsunami runup. In this section, earthquakes are primarily examined, based on the availability of instrumental source data and associated statistical models. It should be noted, at minimum, that seismically triggered submarine landslides can cause anomalous fluctuations in near-field runup, beyond what is expected from earthquake sources, although it has been exceedingly difficult in the past to make a direct connection from a landslide observed in seafloor mapping to a measurable tsunami. In the examples presented below, it is unlikely that there were landslide components to the tsunami.

In the near-field broadside regime, the focus will be on tsunami arrivals using tide gage measurements and tsunami severity related to runup measurements. Tide gage measurements are routine and temporally resolved, whereas runup measurements taken during post-tsunami field surveys are indicative of only the maximum flow depth or runup throughout the inundation of multiple waves arrivals (Farreras, 2000). Runup measurements consist of a variety of water-level measurements made within the inundation zone, including flow depths relative to the local topographic elevation and runup at the point of maximum inundation relative to sea level at the time of the tsunami. Vertical measurements are made from flow indicators or marks left by the tsunami (e.g. vegetation discoloration from salt water and bark stripped from trees). Measurements are made relative to a horizontal datum and must be corrected for the tidal height at the time of the tsunami (Baptista *et al.*, 1993). There is inherent uncertainty in these measurements, owing to misidentification of water marks and tsunami catalogs that do not distinguish runup from flow depths (Borrero, 2001).

The near-field tsunami runup problem is divided into two coastal regimes: broadside runup (i.e. directly across from the rupture zone at strike-normal azimuths) and oblique runup (Fig. 1). This is a logical division inspired by the theoretical work on near-field wave characteristics by Carrier (1995). In that study, he provided an elegant solution to the broadside and oblique runup problem that is expanded from the canonical hodograph transformation of Carrier and Greenspan (1958).

Several working hypotheses are presented below for the near-field broadside regime and throughout the rest of this study for the other two regimes based on both observed tsunami records and our current understanding of tsunami physics.

HYPOTHESIS 1. *For reasonably regular coasts, maximum offshore tsunami amplitude is most often associated with the first arrival, a non-trapped phase.*

This hypothesis is surprisingly difficult to test for several reasons: (1) many near-field tide gage stations become inoperable during significant tsunamis, (2) the definition of the near-field broadside regime is dependent on the length of rupture and spatially restricts which tide gage stations can be used for a test, and (3) there is a difficulty in defining what broadside is for very irregular coasts. An example is shown in Fig. 9a where the Corinto

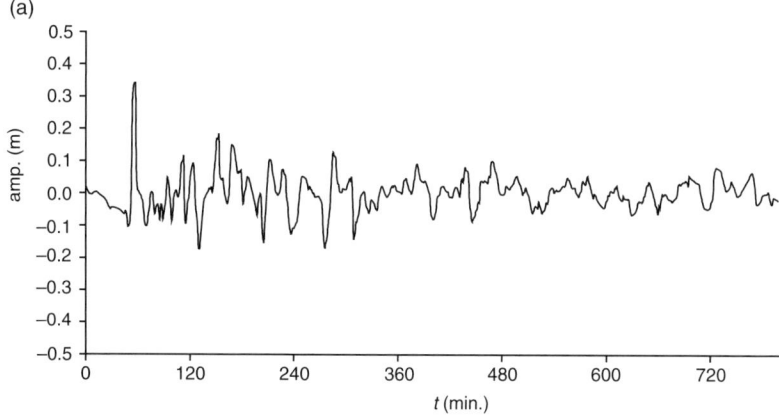

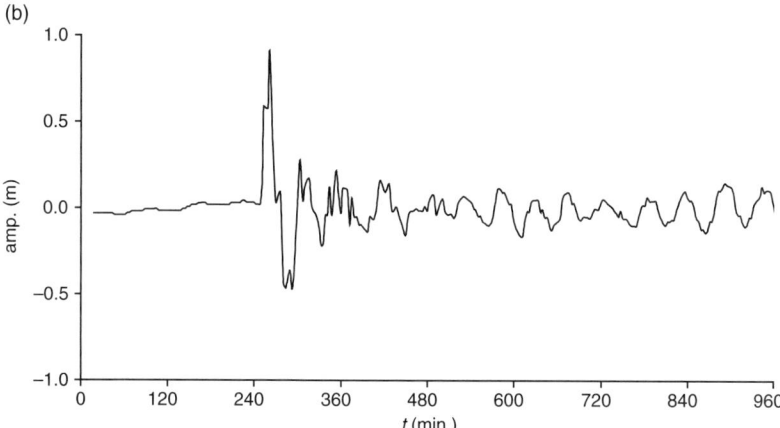

FIG. 9. Examples of near-field and regional tide gage records that are broadside from the rupture zone: (a) Corinto (Nicaragua) record of the 1992 Nicaragua tsunami; (b) Gan (Maldives) record of the 2004 Indian Ocean tsunami.

tide gage station recorded the tsunami from the 1992 Nicaragua tsunami earthquake (Baptista et al., 1993). (The nearby Puerto Sandino tide gage station became inoperable during this tsunami.) The Corinto station clearly shows that the maximum amplitude is associated with the first or direct arrival.

In addition, a few regional tide gage recordings are available for the 2004 Indian Ocean tsunami (Rabinovich and Thomson, 2007) that are in line with the primary beaming pattern from the source region (Geist et al., 2007) and hence could be considered as broadside recordings. For example, the Gan tide gage record from the Maldives shown in Fig. 9b also shows that the maximum amplitude is associated with the first arrival. In general, many island stations have a much less resonant response than continental stations (Watanabe, 1972).

Many Japanese tide gage stations recorded the tsunami from the 2003 Tokachi-Oki earthquake (Hirata *et al.*, 2004), showing the transition of behavior from the near-field broadside to near-field oblique regimes (Fig. 10). The broadside stations include Shoya (which became inoperable) and Tokachikou (which went off scale, interpolated by a dashed line). The Kushiro, Akkeshi, and Krittapu stations are near the source region, but away from the main region of uplift located near the epicenter (star). From the available data, it does appear that the maximum amplitude is associated with the first arrival. While this hypothesis works well for simple coasts, it may not hold for highly irregular coastlines where the maximum amplitude is associated with a later phase (not the direct arrival), as evident from broadside tide gage records from the 1964 Gulf of Alaska tsunami (Lander, 1996).

Analytic studies of runup dynamics provide an understanding of how the offshore waveform relates to runup at the point of maximum inundation. Carrier (1995) uses a Gaussian initial tsunami waveform to show the evolution of tsunami runup with time. Broadside from the source region, the wave history predicted by analytic theory is characterized by a solitary runup/drawdown phase associated with the first arrival of the wave. The exact wave history (i.e. wavelength and amplitude) depends on how far the source region is from shore. Carrier *et al.* (2003) expands the broadside analysis for other waveforms that correspond more to vertical displacement fields associated with elastic dislocations from earthquakes as well as landslide movements. Results from these studies are consistent with Hypothesis 1, although they inherently assume a wave profile that is homogeneous along strike. Therefore, results from actual earthquakes may be restricted to regions broadside from high slip patches on the fault (e.g. see Fig. 10).

HYPOTHESIS 2. *Strike-parallel distribution of maximum amplitude and runup is significantly affected by fault slip heterogeneity.*

In this case, a causal link is proposed between slip heterogeneity and variations in runup distribution (cf. Geist (2002)). As with Hypothesis 1, testing is difficult in regions with irregular shorelines or events along island arcs, for example, which results in scattering and excitation of trapped modes. Again, the 1992 Nicaragua tsunami earthquake provides a good case study, in that there are two distinct slip patches separated by a significant distance, and the coastline and offshore bathymetry are relatively uniform (Fig. 11). The moment distribution is determined from the inversion of seismic surface waves by Ihmlé (1996a,b). The slip distribution is directly related to the moment distribution, assuming a depth-dependent shear modulus (Geist and Bilek, 2001). The runup distribution shown in Fig. 11 is from two separate post-tsunami field surveys (Abe *et al.*, 1993; Baptista *et al.*, 1993). Regions of high runup are directly correlated in a strike-parallel sense with regions of high slip. However, there is significant runup in the intervening regions (broadside from regions of very low slip) caused by the constructive interference of spreading waves from the two regions of high slip (Geist and Dmowska, 1999).

The 1952 and 2003 Tokachi-Oki tsunamis provide another test for Hypothesis 2. In this case, the moment magnitude ($M_W = 8.1$ and $M_W = 8.0$, respectively) and epicenter for the two earthquakes are very similar (Fig. 12). Therefore, difference in the runup distribution can be ascribed primarily to difference in the slip distribution between the

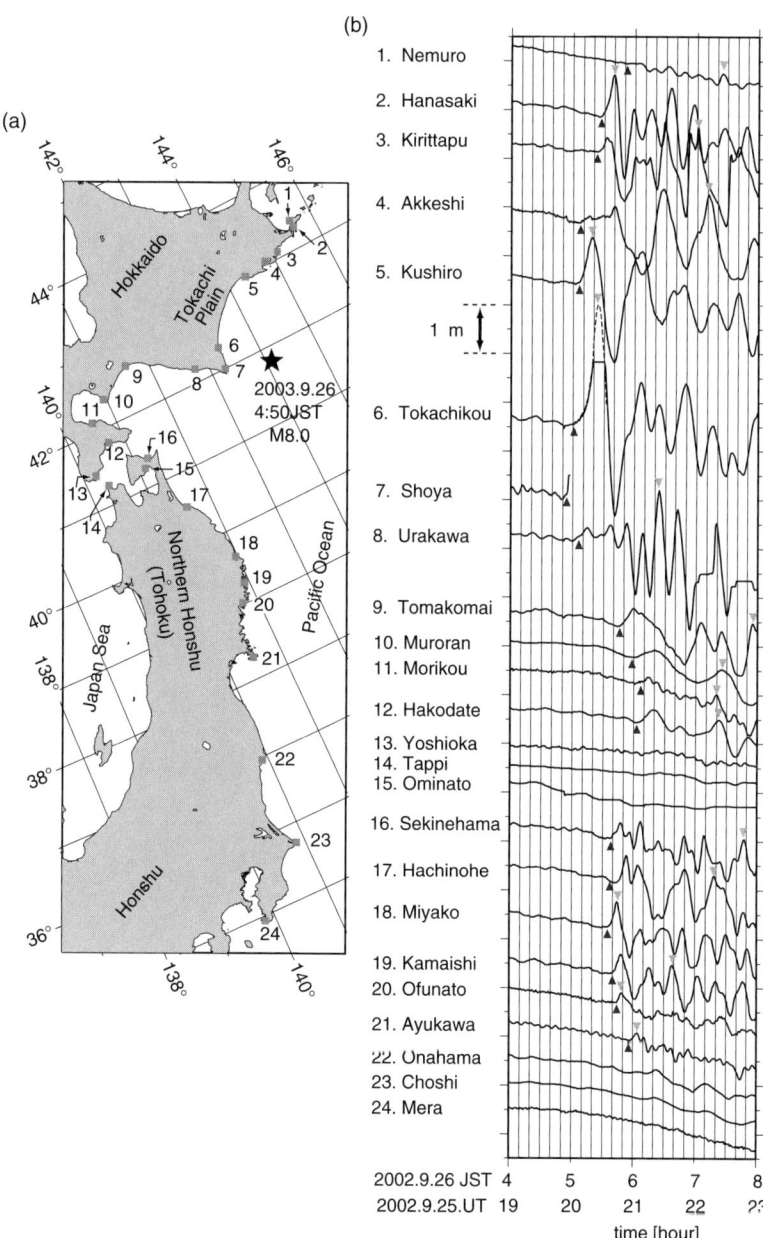

FIG. 10. Near-field Japanese tide gage records of tsunami from the 2003 Tokachi-Oki earthquake (modified from Hirata et al. (2004)). (a) Station location map. Star indicates the location of the earthquake epicenter. (b) Tide gage records (including tidal component); dashed where interpolated. Upward triangle marks the first arrival time; downward triangle marks the time of maximum peak amplitude.

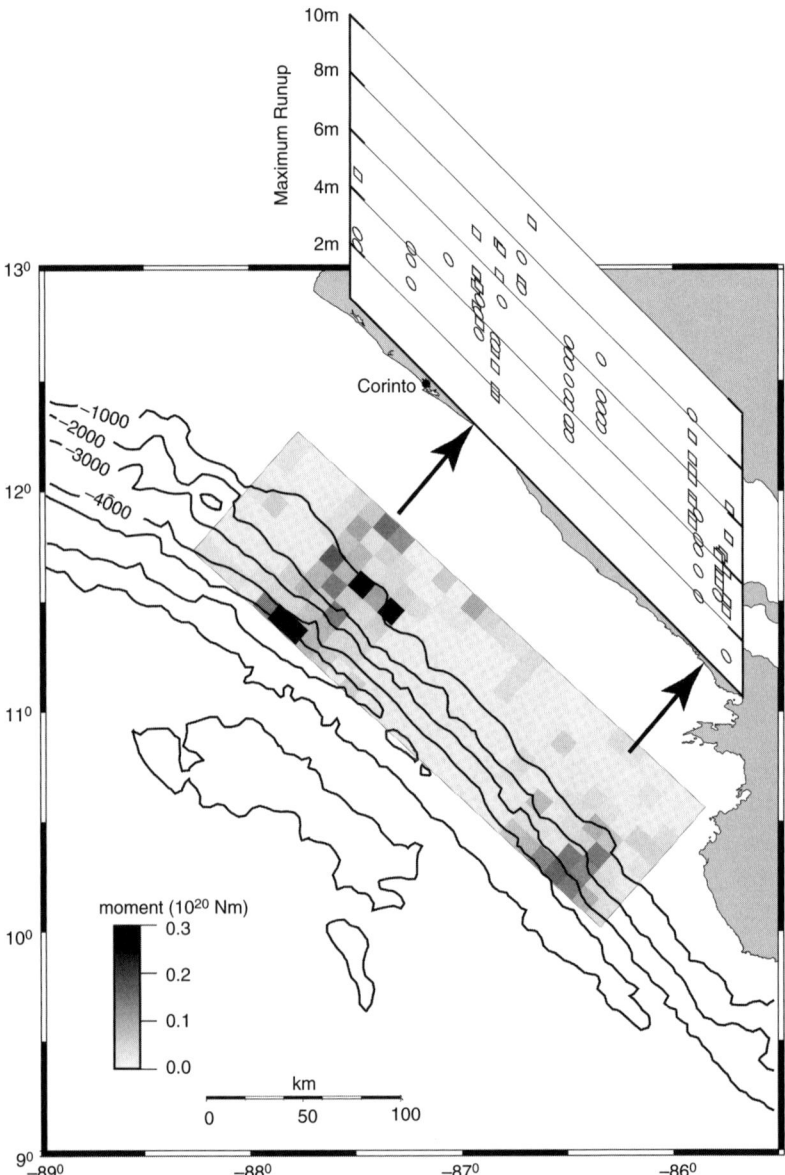

FIG. 11. Correlation of along-strike rupture complexity and along-shore runup for the case of the 1992 Nicaragua tsunami earthquake. Moment distribution from inversion of seismic surface waves (Ihmlé, 1996a,b). Runup distribution from field surveys: Abe *et al.* (1993) (open circles); Baptista *et al.* (1993) (open squares). Bathymetric contour interval: 1000 m.

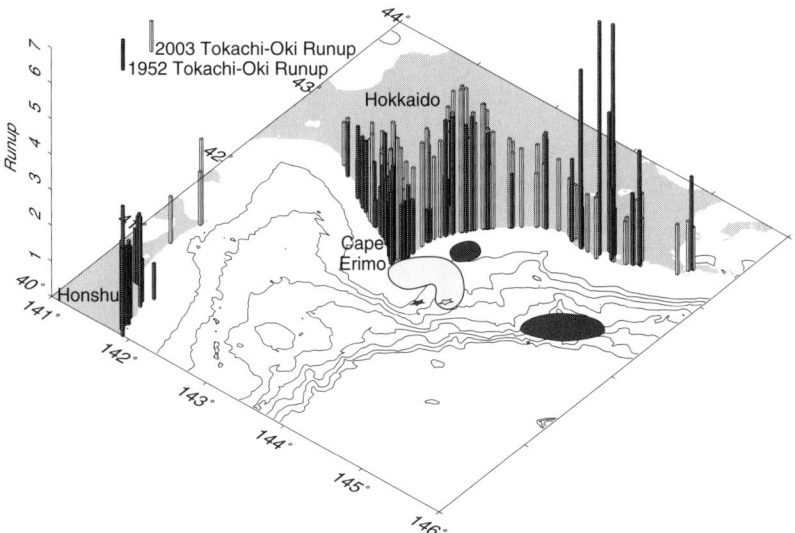

FIG. 12. Runup distribution from tsunami generated by the 1952 (shaded bars) and 2003 (open bars) Tokachi-Oki earthquakes (Tanioka et al., 2004). Epicenters are indicated by shaded and open stars, respectively. Regions of high slip are shown schematically for each earthquake (shaded—1952; open—2003) (cf. Hirata et al. (2003); Yamanaka and Kikuchi (2003)). Bathymetric contour interval: 500 m.

two events. For each case, there are spatially dense runup measurements taken from post-tsunami surveys (Tanioka et al., 2004). The slip distribution for the 1952 earthquake occurred in two regions away from the epicenter: one small region near the Hokkaido coast and another larger region beneath deep water broadside from the NE Hokkaido coast (Hirata et al., 2003). The slip distribution for the 2003 earthquake occurred near the epicenter, toward Cape Erimo and primarily beneath shallow water (Yamanaka and Kikuchi, 2003). As with the 1992 Nicaragua tsunami, regions of high runup are correlated along shore with regions of high slip along strike (Fig. 12).

More quantitative descriptions of runup observations are provided by statistical models. Existing statistical models for the spatial distribution of runup (R) are based on the log-normal distribution. In particular, Choi et al. (2002) examined measured runup values in the near-field (not distinguishing between broadside and oblique regimes) for 11 tsunamis in the Pacific from 1992 and 1998 and found that all of the events, except for one, can be fit by a log-normal distribution with a density distribution

$$f(R) = \frac{1}{R\sigma\sqrt{2\pi}} e^{-\frac{(\ln(R)-\mu)^2}{2\sigma^2}}, \qquad (24)$$

where μ and σ are the mean and the standard deviation of $\ln(R)$, respectively. The exception is the 1996 Sulawesi tsunami that is best fit by a power-law distribution. In a follow-up study, Choi et al. (2006) examined the runup distributions associated with the 2004 Indian Ocean tsunami in eight different regions and again concluded that a log-normal distribution is the optimal statistical model.

Although one-point statistical models, such as the log-normal model described above, adequately fit the observed data, one would expect that there should be a characteristic length-scale of broadside runup that is proportional to the fault rupture length (along strike). Okal and Synolakis (2004) attempted to fit the maximum runup distribution parallel to the near-field shoreline (y-axis) for 9 tsunamis (spanning the years 1946-2002) with the following shape function:

$$R(y) = \frac{b}{[(y-c)/a]^2 + 1}, \qquad (25)$$

where a, b, and c are adjustable parameters. They found that the aspect ratio $I_2 = b/a$ scales with the size of the earthquake and that I_2 for runup associated with near-field landslide tsunamis is distinctly different from that for seismogenic tsunamis. The logical next step of developing empirical two-point statistical models of tsunami runup is difficult, owing to the uncertainty and non-uniformity of maximum runup measurements in the field. Alternatively, deriving statistical runup models from two-point slip models described in Section 3.1 is equally difficult, owing to the complexity imposed by wave propagation and the runup process itself (i.e. "site response"). An analogous connection between statistical slip models and seismic ground motions, with its inherent seismic propagation and site response complexities, has been investigated by many researchers (e.g. see Tsai (1997b); Berge *et al.* (1998); Somerville *et al.* (1999); Guatteri *et al.* (2004)). Presently, therefore, the optimal model for runup distribution may indeed be the log-normal distribution discussed by Choi *et al.* (2002), especially when considering the complexity of wave propagation and response in the near-field regime and the possibility of triggered landslide tsunamis contributing to the total tsunami wavefield.

HYPOTHESIS 3. *Broadside runup increases where high coseismic slip is located beneath deep water.*

Near-field broadside runup is not only dependent on the magnitude of vertical seafloor displacement derived from slip on the fault, but also the water depth where this displacement occurs. As a tsunami propagates from deep water to shallow water, the amplitude increases according to Green's law derived from ray theory or the conservation of energy flux:

$$A \propto b^{-1/2} h^{-1/4}, \qquad (26)$$

where b is the distance between rays and h is the water depth. Typically, the variation in b is negligible so that Green's law is reduced to $A \propto h^{-1/4}$. Although the quarter exponent suggests a weak dependence with depth, ocean depths where tsunamis are generated (from shelf to trench) can vary by at least two orders of magnitude, making the shoaling amplification effect significant. In addition, the wavelength shortens and leading wave steepness increases during shoaling, as discussed by Mei (1989). The wavelength, height and steepness, in addition to leading polarity, are all important factors in determining runup (e.g. see Pelinovsky and Mazova (1992); Tadepalli and Synolakis (1994); Carrier *et al.* (2003); Tinti and Tonini (2005)). For a given amount of slip on the up-dip portion of

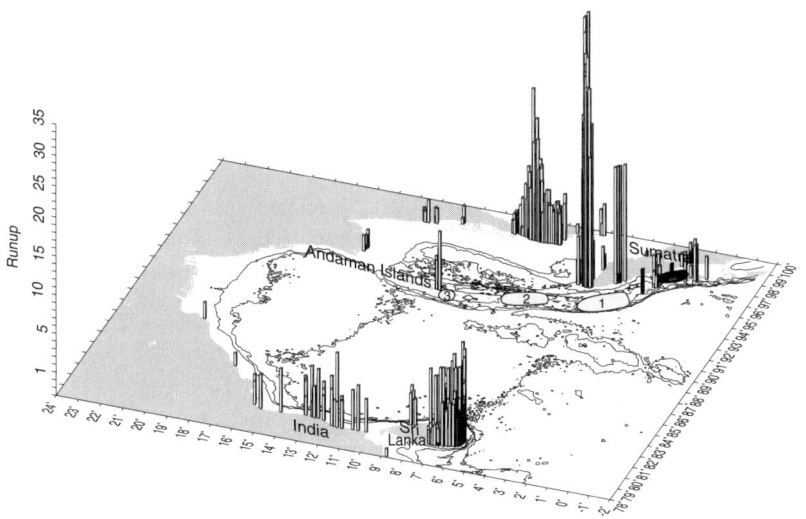

FIG. 13. Runup distribution from tsunamis generated by the 2004 Sumatra-Andaman (open bars) and 2005 Nias (black bars) earthquakes. Epicenters are indicated by open and shaded stars, respectively. Regions of high slip are shown schematically for each earthquake (open—2004; black—2005) (cf. Ammon et al. (2005); Banerjee et al. (2007); Chlieh et al. (2007); Rhie et al. (2007)). Bathymetric contour interval: 1000 m.

the inter-plate thrust, the amplitude and wavenumber of the initial tsunami wave profile will first be higher compared to the same amount of slip on a down-dip portion of the thrust, owing to the effects of the elastic response in the solid earth (see Fig. 7). This difference will then be accentuated during propagation toward shore through the effects of Green's law in the ocean.

This effect can be observed by again examining the comparison between the 1952 and 2003 Tokachi-Oki earthquakes (Fig. 12). The overall maximum runup of the 1952 earthquake is located broadside from a region of high slip beneath deep water. The region of high slip associated with the 2003 earthquake is in shallower water and does not result in maximum runup as large as in 1952, even though the seismic moments for the two earthquakes are very similar.

Another illustrative example is the tsunami runup from the $M_w = 9.2$ 2004 Sumatra-Andaman earthquake (Fig. 13). The slip pattern during this 1200-1600 km long rupture (Ishii et al., 2005; Meltzner et al., 2006; Subarya et al., 2006) was complex and highly variable (cf. Ammon et al. (2005); Banerjee et al. (2007); Chlieh et al. (2007); Rhie et al. (2007)). High slip occurred in primarily three regions (numbered from south to north in the direction of rupture propagation; see Fig. 13). Slip was highest in region 1 and there are suggestions that slip was concentrated on the shallow portion of the inter-plate thrust beneath deep water (Bilek, 2007; Rhie et al., 2007; Seno and Hirata, 2007). Correspondingly, tsunami runup was highest broadside from region 1 (i.e. in NW Sumatra and Sri Lanka). High slip in region 2 is broadside from high tsunami runup across the Andaman Sea in Thailand (e.g. Phuket). Slip in region 3 is difficult to resolve

(Ammon et al., 2005), but is possibly correlated with the runup observed on the Andaman Islands.

The effect of regions of high slip located in deep water has also been discussed in comparing the tsunamis from the 2004 Sumatra-Andaman earthquake and the $M_w = 8.7$ March 28, 2005 Nias earthquake as described by Geist et al. (2006). The maximum runup from the 2004 earthquake was over 30 m, whereas the maximum runup from the 2005 earthquake was approximately 4 m. Although the difference can partly be ascribed to the overall higher seismic moment for the 2004 earthquake, the high slip region broadside from the area of maximum runup possibly occurred in deeper water near the trench in 2004 as mentioned above, whereas high slip occurred either beneath shallow water or on land (Nias and Sumatra Islands) in 2005 (Fig. 13).

3.4. Tsunami Observations—Near-field Oblique Regime

For the near-field oblique case (Fig. 1), observations indicate that the maximum tsunami arrival is often related to edge waves. Two hypotheses can be formulated from the available data and our understanding of tsunami physics.

HYPOTHESIS 4. *Maximum amplitude and runup are most often derived from late arrivals resulting from the interaction of trapped phases (i.e. edge waves).*

A good case study that tests this hypothesis is the tsunami generated from the 1983 $M_w = 7.7$ Nihonkai-Chubu (Sea of Japan) earthquake (Abe and Ishii, 1987). Because edge waves are scattered by longshore irregularities, it is often difficult to track the propagation of edge wave phases in tide gage records. However, Abe and Ishii (1987) show a smooth section of the western Honshu coastline, approximately 160 km long, in which the arrival time of the maximum amplitude is closely aligned with the predicted group velocity of edge waves. Another near-field oblique example that is consistent with this hypothesis is the tsunami record at the Crescent City tide gage station following the 1992 $M_w = 7.2$ Cape Mendocino earthquake (González et al., 1995).

An Airy phase exists at a minimum in the group velocity for edge waves and has been observed in Japan from the tsunami generated by the 1952 $M = 9.0$ Kamchatka earthquake (Ishii and Abe, 1980; Golovachev et al., 1992). The Airy phase is particularly pronounced owing to interference effects, as is commonly observed with Airy phases associated with solid-earth seismic waves (Lay and Wallace, 1995). A minimum in the group velocity for fundamental mode edge waves is predicted from the step bathymetry model (Ishii and Abe, 1980). Although there is an exponential decay associated with scattering of edge waves, including the Airy phase, from an irregular coastline, the e-folding distance can be particularly long for continents. For the NE Japan margin, the e-folding distance is estimated to be over 400 km (Fuller and Mysak, 1977).

The oblique stations of the 2003 Tokachi-Oki earthquake also provide a test to this hypothesis (Fig. 10). For most of the oblique stations such as Urakawa (station 8), the largest wave arrives significantly later than the first arrival (Tanioka et al., 2004). Note, however, the direct propagation path from the earthquake to Miyako (station 18, Fig. 10). The largest wave in this case is the first arrival. As the tsunami proceeds southward down the Honshu coastline, edge waves appear to be dominant relative to the first arrival.

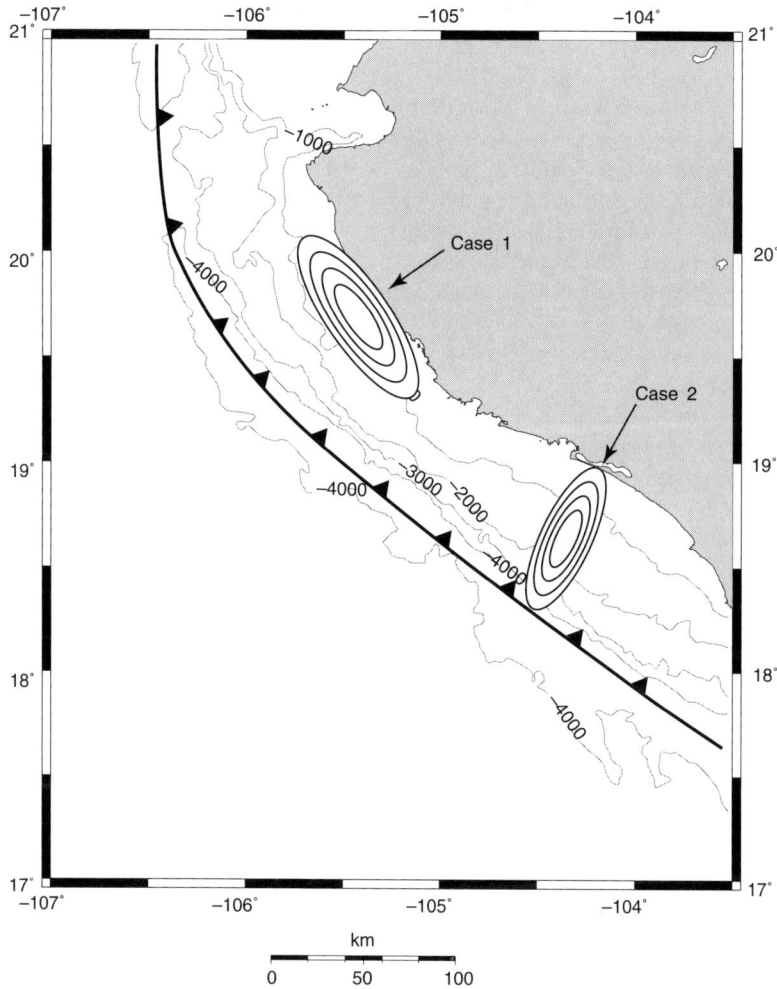

FIG. 14. Two cases of edge wave excitation examined by Fujima et al. (2000) for an initial vertical disturbance near the coastline. Case 1: elliptical disturbance oriented parallel to the coastline (e.g. earthquake). Case 2: elliptical disturbance oriented normal (e.g. landslide).

HYPOTHESIS 5. *There exists a causal (but complex) relationship between fault slip heterogeneity and near-field oblique tsunami amplitude.*

Because edge waves frequently scatter and resonate during propagation in the near-field oblique regime, it is difficult to directly trace the influence that source heterogeneity has on edge wave characteristics. The analytic investigation of edge waves propagating along a linear coastline with uniform slope by Fujima et al. (2000) is heuristically useful in this regard (Fig. 14). They examine what effect the orientation of the initial tsunami wavefield (as specified by the long axis of an elliptical disturbance) has on edge wave characteristics. For the case where the initial disturbance is located near the coast with the long axis oriented parallel to the coast, the fundamental mode edge wave ($n = 0$, Fig. 4) dominates (see also Liu et al. (1998)). This case is apropos for typical inter-plate thrust earthquakes, in which the vertical displacement field spans across and is oriented subparallel with the coastlines (e.g. see Fig. 15). Details of the displacement field which include localized areas of uplift and subsidence appear to have a significant effect on the types of edge waves excited, with higher-mode edge waves associated with components of the displacement field located farther offshore (Kajiura, 1972; Fujima et al., 2000). Kajiura (1972) also demonstrated that edge wave excitation increases with a decrease in the along-strike length of components of the vertical displacement field. A complex displacement field as shown in Fig. 15 is likely to generate many edge wave phases at fundamental and higher modes.

In the second case examined by Fujima et al. (2000), the near-coast displacement is oriented with the long axis normal to the coastline (Fig. 14). In this case, the excited edge wave is dominated by the first-order mode ($n = 1$; see Fig. 4) and is much more irregular than for the first case examined by Fujima et al. (2000). This case is apropos for nearshore landslides. Lynett and Liu (2005) reached similar model-based conclusions specific to landslide tsunamis. Both Fujima et al. (2000) and Lynett and Liu (2005) noted the possibility that the maximum amplitude associated with the near-field oblique wave can be greater than that for the near-field broadside wave. For an initial disturbance located far offshore (e.g. as with slow tsunami earthquakes located near the trench), little edge wave energy is excited for the ideal case of a uniform coastline (Kajiura, 1972; Fujima et al., 2000). However, slight perturbations at the coast can also excite edge waves (Guza and Davis, 1974; Fuller and Mysak, 1977).

4. Far-field Regime

In this section, time-series statistics from deep-ocean and shore-based measurements are examined in the far field. The dominant aspect of the far-field tsunami is the tsunami coda evident in time-series records. Whereas the leading wave from the source tends to retain its initial waveform, the maximum tsunami amplitude at the coast most often occurs during the arrival of the coda. It is difficult to accurately model the tsunami coda, both in the open ocean and at the coast, using numerical models. The coda observed in the deep ocean and at the coast can be described from a phenomenological perspective, focusing on the amplitude and timing of the largest observed wave. Tsunami bottom pressure recorder (BPR) and tide gage data from the recent November 2006 and January 2007 Kuril Islands earthquakes are used as an example. Figure 16 shows the locations of the earthquakes and the stations where tsunami time-series data are discussed. The data shown in Fig. 16 also include a seismic surface wave arrival (under-sampled at a sampling rate of 1 minute) that precedes the tsunami direct arrival and the tsunami coda. The seismic surface wave is stronger for the January 2007 outer-rise normal faulting earthquake compared to the November 2006 inter-plate thrust earthquake because of differences in the seismic radiation patterns.

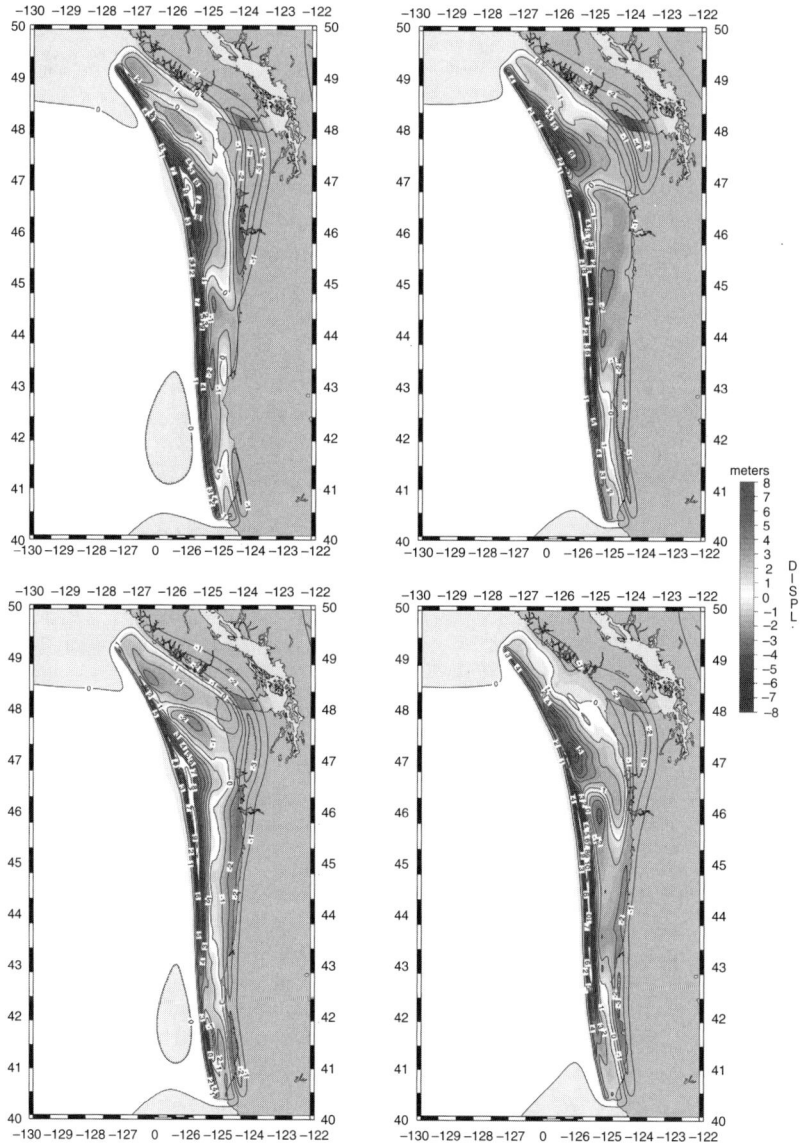

FIG. 15. Four examples of possible vertical coseismic displacement fields from slip on the Cascadia inter-plate thrust, showing a predominantly subparallel orientation of the axis of displacement with the coastline. Displacement is calculated using the static elastic displacement equations (Eqs (3) and (4)) with fractal slip (Eq. (17)) and a non-planar fault geometry (Flück et al., 1997).

140 GEIST

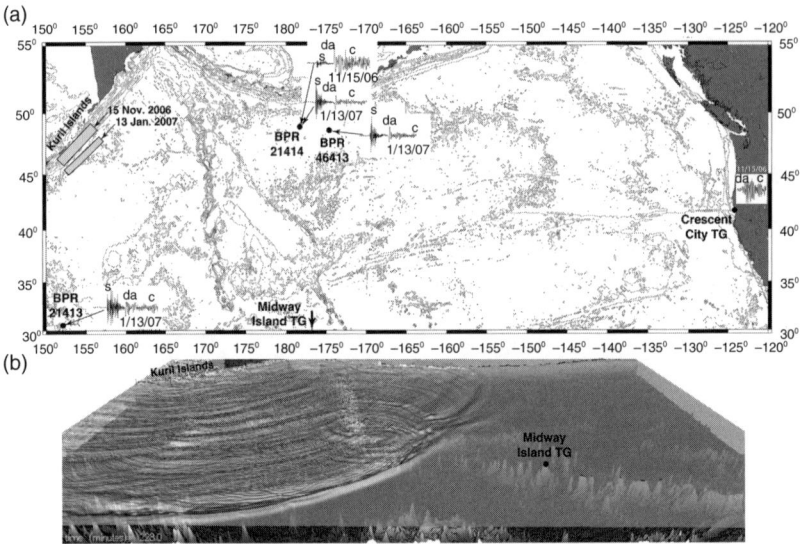

FIG. 16. (a) North Pacific map showing 15 November 2006 and 13 January 2007 Kuril earthquakes (light gray rectangles) and time-series data locations (filled circles). BPR: bottom pressure recorder. TG: tide gage. Sample detided time series indicated for both tsunamis. Arrivals: s-seismic surface waves (under-sampled), da-tsunami direct arrival, and c-tsunami coda. Bathymetric contour interval: 1 km. (b) Snapshot at 228 min. of computed tsunami wavefield from the 2006 Kuril earthquake. Perspective view to the north.

4.1. Open-ocean Propagation

HYPOTHESIS 6. *The deep-ocean far-field tsunami wavefield includes a well-developed coda caused by frequency dispersion, scattering, reflected, and refracted arrivals.*

During propagation, the tsunami wavefield gradually evolves, owing to the effects of dispersion and of reflection, scattering, and focusing from changes in bathymetry. As an example of this complexity, a snapshot of the numerically simulated tsunami wavefield in the north Pacific from the November 2006 earthquake is shown in Fig. 16b. In this section, I first examine the coda recorded from BPR time-series measurements both in terms of its amplitude distribution and its spectral content. I then review possible origins for the coda, including dispersion and scattering.

The evolution of far-field, deep-ocean waveforms has been discussed since the advent of bottom pressure recorders, particularly with regard to the effects of dispersion (e.g. see González and Kulikov (1993)). The recent 2006 and 2007 Kuril tsunamis are useful case studies since there are several available observation points along unobstructed ray paths (Fig. 16). Time series of tsunami water-level elevations (converted from pressure measurements) at these observation points are plotted in Fig. 17. It is apparent in these time series that the direct arrival phase is stable, while the coda of the time series is highly variable. The direct arrival phase, in accordance with the principle of causality, follows a least-time ray path. The coda, in contrast, is described by the combined effects of dispersion and scattering as described below.

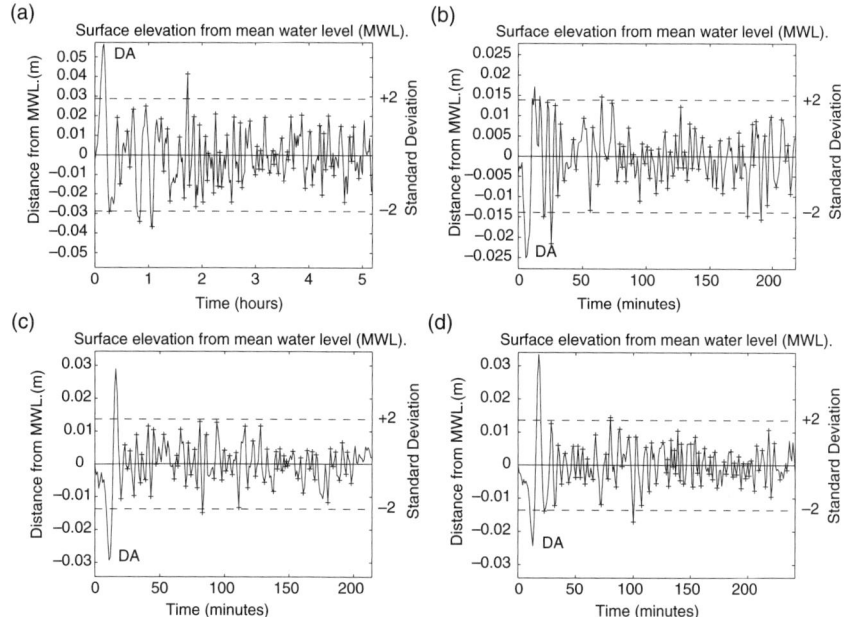

FIG. 17. Time series of tsunami bottom pressure measurements from locations indicated in Fig. 16, showing the direct arrival (DA) and the ensuing tsunami coda. Location of crest and trough amplitudes is indicated by + symbol. (a) 15 November 2006 Kuril tsunami, station 46413. (b) 13 January 2007 Kuril tsunami, station 21413. (c) 13 January 2007 Kuril tsunami, station 21414. (d) 13 January 2007 Kuril tsunami, station 46413.

The coda is typically characterized by an exponentially decaying wave envelope. Much of the analysis and theory of the seismic coda can be directly adapted to analyze the tsunami coda: a review of past research related to the seismic coda is discussed in Volume 50 of Advances in Geophysics. The envelope of tsunami amplitudes is assumed to follow an exponential decay (Mofjeld et al., 2000a):

$$A_c = A_0 \sigma \exp\left[-\left(t - t_0\right)/\tau\right], \qquad (27)$$

where A_0 is a constant coefficient, σ is the initial standard deviation of A_c, and τ is the e-folding decay constant. This expression is analogous to a frequency-independent (cf. van Dorn (1984)) form of energy density for the seismic coda (Aki and Chouet, 1975)), where τ is analogous to the seismic attenuation factor Q. To be consistent with the physics of the origin of the coda, I define it as commencing immediately after the first arrival, whereas Mofjeld et al. (2000a) chose $t_0 = 4$ hours for tsunami forecasting objectives. To estimate the coda envelope, I compute a running root mean square of the water-level elevation η (Nishigami, 1991; Nishigami and Matsumoto, 2008) and find the best-fit exponential curve. Alternatively, Mofjeld et al. (2000a) provided a probabilistic definition of the tsunami coda envelope based on a cumulative binomial distribution.

The initial statistical model of water-level elevations (η) within the coda is that of a time-varying Gaussian distribution (Takahara and Yomogida, 1992). The

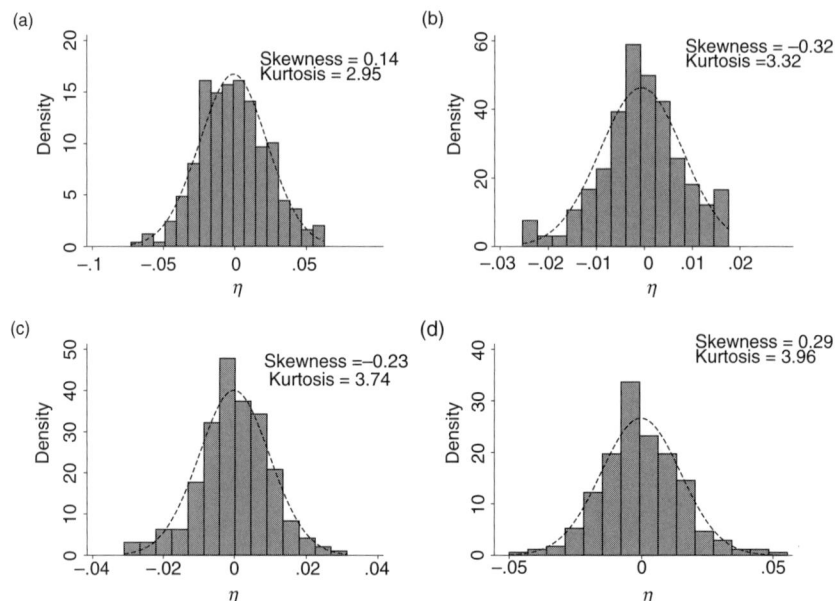

FIG. 18. Empirical density distribution of tsunami coda waves from data plotted in Fig. 17. Dashed line indicates the Gaussian distribution model.

stationary Gaussian distribution is also the canonical model for deep-water waves in a fully developed random sea (e.g. see Longuet-Higgins (1952)), though Gaussian transformations (Rychlik et al., 1997) and second-order statistical models have been proposed that more accurately reflect real wind-generated waves in the ocean (e.g. see Hogben (1990); Huang et al. (1990); Jha and Winterstein (2000); Prevosto et al. (2000); Muraleedharan et al. (2007)). To determine whether tsunami coda waves conform to a Gaussian distribution, the detided data are first corrected for the exponential envelope decay (Eq. (27)) to simulate a stationary wave sequence, and an empirical density function is then determined (Fig. 18). The Shapiro-Wilk test for normality (Shapiro and Wilk, 1965; Stephens, 1974) indicates that Gaussian null hypothesis cannot be rejected for two of the stations (Fig. 18a and c) but can be rejected at 95% confidence for the other two stations (Fig. 18b and d).

The distribution of amplitudes corresponding to the Gaussian distribution is a Rayleigh distribution (Rayleigh, 1880). Longuet-Higgins (1952) derived this correspondence for the case of a narrow band of frequencies for ocean waves. Spectra for deep-ocean tsunami waves are shown in Fig. 19, indicating the frequency band where the tsunami energy occurs. The Rayleigh density distribution is typically given in terms of the crest-to-trough wave height (H):

$$f(H) = \frac{2H}{H_{rms}^2} \exp\left(-\frac{H^2}{H_{rms}^2}\right), \tag{28}$$

where H_{rms} is the root-mean-square wave height. For amplitudes, the distribution can be

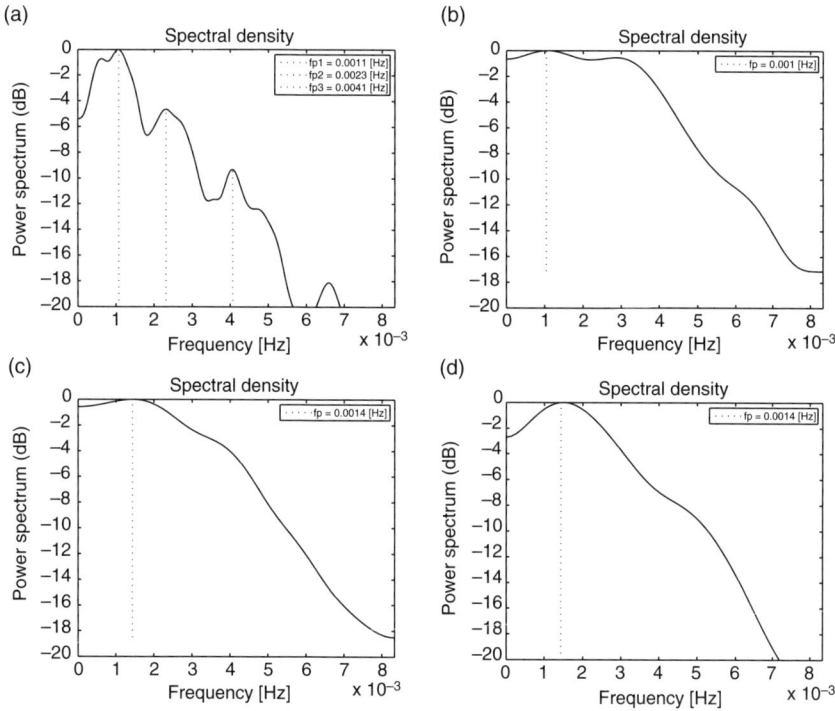

FIG. 19. Spectra of tsunami coda waves from data plotted in Fig. 17.

written as (Massel, 1996):

$$f(A) = \frac{A}{\sigma_\eta^2} \exp\left(-\frac{A^2}{2\sigma_\eta^2}\right), \tag{29}$$

where σ_η^2 is the variance of the water-level elevation time series. For a stochastic time series, definitions of wave amplitude and wave height are not necessarily straightforward. Amplitude can be defined as $A = \frac{1}{2}H$ (Longuet-Higgins, 1952) or the crest (trough) amplitude $A_c(A_t)$ can be defined as the global maximum (minimum) between successive downcrossings (Rychlik and Leadbetter, 1997) as shown in Fig. 17. In this case, $H = A_c - A_t$. For transient waves such as tsunamis, the crossings refer to the ambient, mean sea level prior to the onset of the tsunami. It should be noted that multiple local maxima and minima, defined by turning points, can exist between successive downcrossings. Statistics of these smaller oscillations are mainly of interest for fatigue analysis in wave engineering (Rychlik and Leadbetter, 1997). Shown in Fig. 20 is the empirical cumulative distribution of wave height in comparison to the corresponding Rayleigh distribution. Including the direct arrival wave height can result in a poorer fit when it is much greater than the tsunami coda wave heights (e.g. see Fig. 20d).

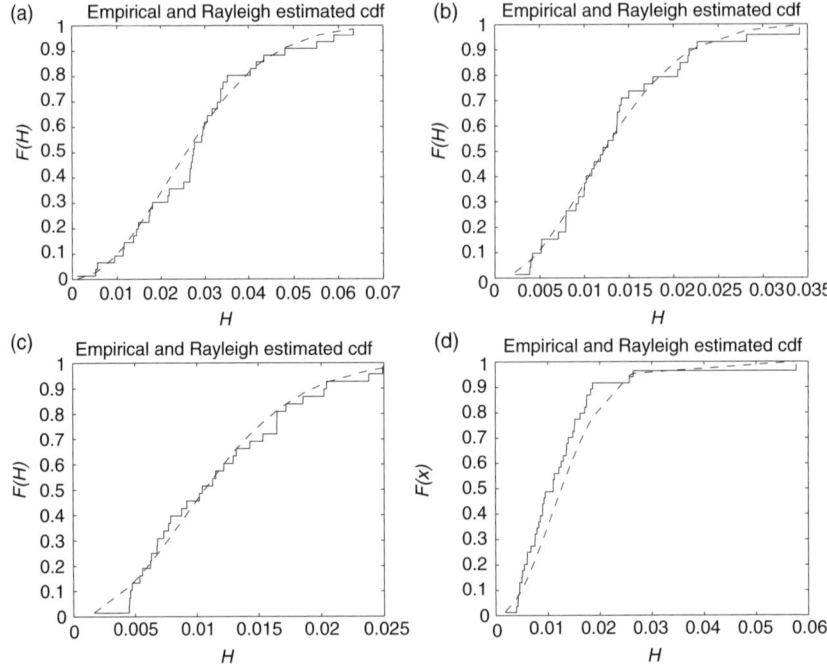

FIG. 20. Empirical cumulative distribution (solid line) of wave height (H) from data shown in Fig. 17. Dashed line indicates the Rayleigh cumulative distribution model.

Given the statistical properties of deep-ocean tsunami coda waves described above, it is instructive to examine analytic functions of dispersion and scattering derived by several authors, as described below. Analytic theory shows the evolution of the leading wave of the tsunami over large distances and the development of the dispersive coda. The first cycle of the leading wave is generally stable, allowing source characteristics to be inferred from the inversion of far-field measurements (e.g. see Fujii and Satake (2008)). To develop analytic expressions of the evolving tsunami waveform, several prescribed expressions of initial, static seafloor displacement have been considered. These are displayed in Fig. 21 along with an example from elastic displacement theory (Eqs (3) and (4)) and a spatially heterogeneous source.

In one dimension and for a constant water depth, the evolution of a linear dispersive wave from a monopole initial disturbance can be derived using the stationary-phase approximation (Jeffreys and Jeffreys, 1956; Stoker, 1957). The equation for the leading part of the wave is derived by Wu (1981) and Mei (1989) as:

$$\eta(x,t) = \frac{m}{2\rho} (\gamma t)^{-1/3} Ai \left[\frac{x - ct}{(\gamma t)^{1/3}} \right], \qquad (30)$$

where m is the excess mass of the initial wave, $c = (gh)^{1/2}$, $\gamma = \frac{1}{6}ch^2$ and Ai is the Airy function. Carrier (1971) provided a specific expression for the case of a Gaussian initial

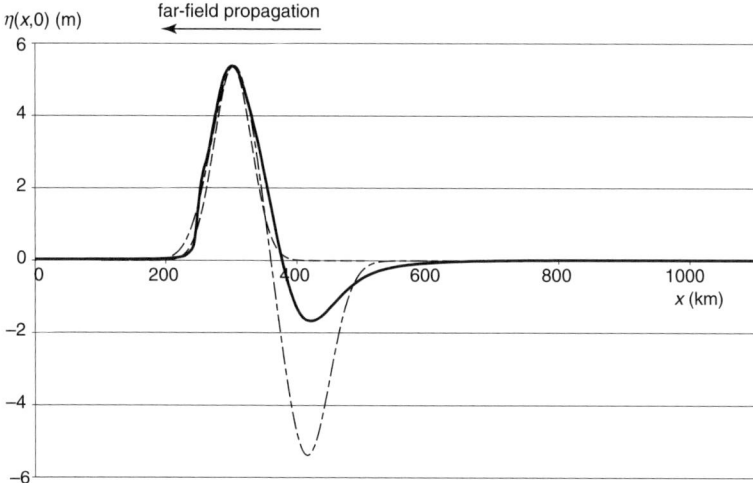

FIG. 21. Monopole (dashed) and dipole (dot-dash) analytic functions of vertical surface displacement used to specify the initial conditions for tsunami propagation. Shown for comparison is vertical displacement computed surface displacement for the 2004 Sumatra-Andaman earthquake at the latitude of Great Nicobar Island, using Chlieh et al.'s (2007) slip model (heavy solid line).

disturbance (Fig. 22a) and Kajiura (1963) provided expressions for other prescribed initial disturbance functions. In each case, the amplitude of the leading wave decays as a function of $t^{-1/3}$. For a dipole disturbance (Fig. 21) involving tilting of the seafloor and zero excess mass, the corresponding wave equation is (Wu, 1981; Mei, 1989)

$$\eta(x,t) = \frac{l^2}{2h} (\gamma t)^{-2/3} Ai' \left[\frac{x - ct}{(\gamma t)^{1/3}} \right], \qquad (31)$$

where l is the dipole moment and $Ai'(z) \equiv \frac{d}{dz} Ai(z)$ (Fig. 22c). In this case, the amplitude of the leading wave decays more rapidly as a function of $t^{-2/3}$ (Fig. 22d).

For 2D propagation in constant water depth, the expressions are more complex. Tsunami radiation is particularly sensitive to fault strike (Gica et al., 2007), with energy primarily focused along azimuths normal to strike. Longer rupture lengths produce a greater beaming effect along these azimuths (Ward, 1982). For a simple elliptic initial displacement in an ocean of constant depth, Kajiura (1970) indicates that the radiation pattern as a function of azimuth (ϕ) is given by

$$R(\phi) = \left[\cos^2 \phi + (b/a)^2 \sin^2 \phi \right]^{-1}, \qquad (32)$$

where a and b are the lengths of the major and minor axes. Other studies describe a more complex radiation pattern for continental shelf bathymetry (Kajiura, 1972) and in direct relation to earthquake source parameters (e.g. see Ben-Menahem and Rosenman (1972); Ward (1982); Okal (2003)). In particular, the finiteness factor of an earthquake source

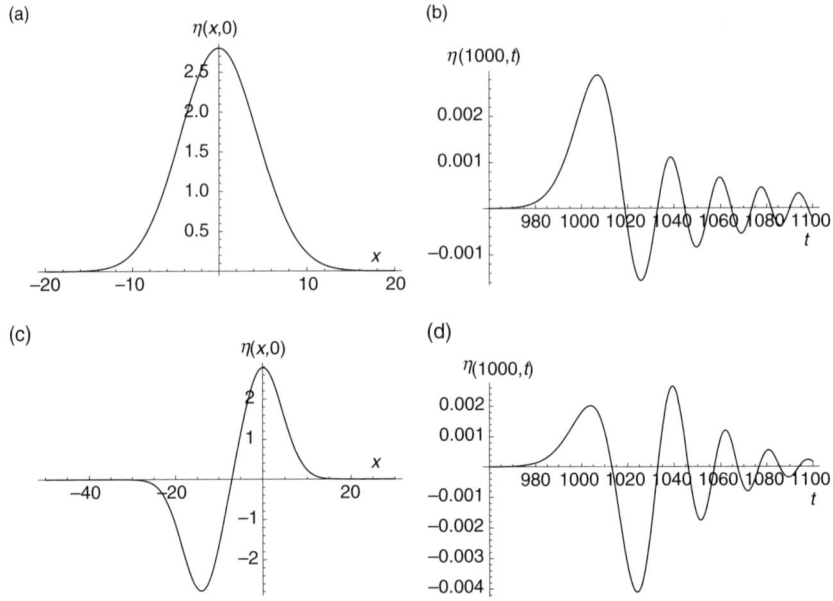

FIG. 22. For the monopole source shown in (a), the far-field time series is shown in (b). For comparison, a dipole source (c) is used to generate the far-field time series shown in (d) at the same distance (Carrier, 1971). Non-dimensional units are used.

takes the form

$$\frac{\sin X}{X}, \quad \text{where } X = \frac{1}{2}kL\left(\frac{c}{v_r} - \cos\phi\right), \tag{33}$$

which includes directivity caused by rupture propagation with speed v_r (Ben-Menahem and Rosenman, 1972; Okal, 1988). Because for tsunamis, $c \ll v_r$, the latter effect is typically small, but can be noticeable for long ruptures such as the 2004 Sumatra-Andaman earthquake (e.g. see Geist et al. (2007)). Figure 23 shows the radiation pattern including these effects. For elongated ruptures (Fig. 23b), note the rotation of the radiation pattern in the direction of rupture propagation and the presence of multiple, smaller lobes.

In place of the Airy wave function for the 1D case, the following function is defined by Kajiura (1963) that dictates the shape of the waveform in two horizontal dimensions:

$$T(p) = \text{Re}\left[(1+i)\int_0^\infty e^{i(u^2 p + u^6)} du\right], \tag{34}$$

where

$$u^6 = \frac{(kh)^3}{g}\left(\frac{g}{h}\right)^{1/2} t \tag{35}$$

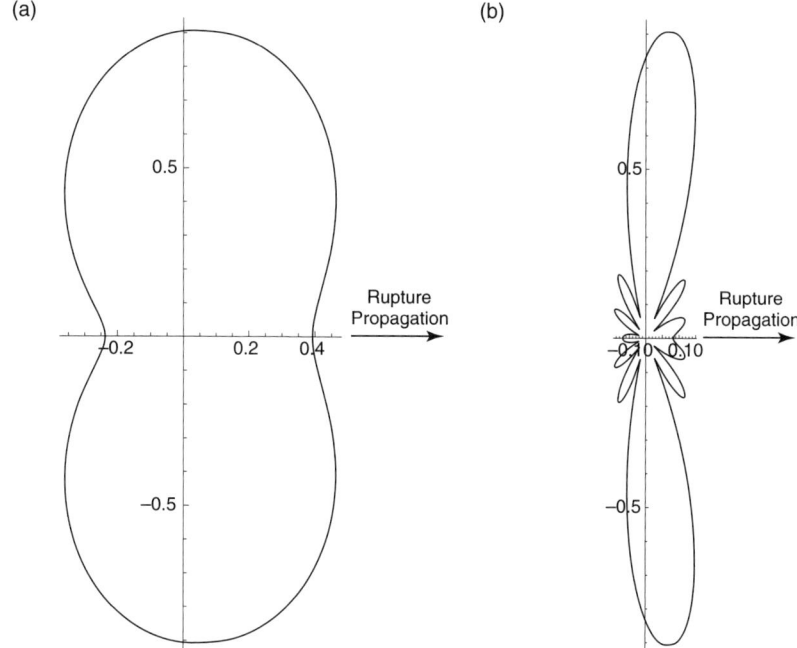

FIG. 23. Tsunami radiation pattern for a kinematic model of earthquake rupture in a constant depth ocean (Ben-Menahem and Rosenman, 1972). (a) 3:1 aspect ratio of source dimensions for typical tsunamigenic, inter-plate thrust earthquakes; (b) 12:1 aspect ratio for an elongated rupture.

and

$$p = \frac{r/h - (g/h)^{1/2} t}{\left[(g/h)^{1/2} t/6\right]^{1/3}}. \tag{36}$$

For a 2D monopole displacement, $\eta \sim -T_p$, where $T_p \equiv \frac{d}{dp} T(p)$. Mei (1989) examines a specific case for a 2D dipole (antisymmetric) displacement where

$$\eta(r, \theta, t = 0) = \frac{A}{a} \left(a^2 - r^2\right)^{1/2} \cos(\theta). \tag{37}$$

Using the stationary-phase approximation, Mei (1989) derived the following expression for the leading wave emanating from this displacement pattern.

$$\eta(r, \theta, t) = \cos\theta \frac{Aa^3}{16 (2r)^{1/2}} \frac{T_{pp}}{h^{5/2} \left[(gh)^{1/2} t/6\right]^{5/6}}. \tag{38}$$

The amplitude for the 2D dipole case decays as a function of $t^{-4/3}$.

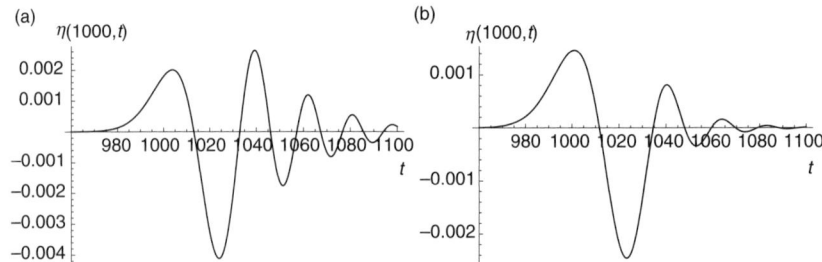

FIG. 24. Effect of source width on far-field dispersion using the Carrier (1971) dipole source (Fig. 22c). (a) $\alpha = 10$, (b) $\alpha = 30$.

To examine the effect that dispersion has on the coda, Carrier (1971) used the following Gaussian function of width α

$$\eta(x, y, t = 0) = \frac{1}{2} \frac{e^{-x^2/4\alpha}}{(\pi\alpha)^{1/2}}. \tag{39}$$

The corresponding far-field tsunami waveform near the leading part of the wavetrain is given in the non-dimensional form as

$$\eta(x, y, t) = \frac{1}{2} (2/t)^{1/3} \, Ai \left[x - t + \left(2\alpha^2/t\right) \right] e^{[(8\alpha^3/3t^2) - 2\alpha(t-x)/t]} \tag{40}$$

(cf. Eq. (34) above). Carrier (1971) demonstrated that dispersion increases with a decrease in the width parameter α, corresponding to a decrease in the down-dip width of the rupture zone ($\alpha \sim W \cos(\delta)$) (Fig. 24). Moreover, Carrier (1971) approximated a dipole wave with the function (Fig. 21)

$$\eta_{\text{dipole}}(x, y, t) = \eta(x, y, t) - \eta(x + b, y, t). \tag{41}$$

For a narrow source, the leading far-field waveform is characterized by a trailing peak amplitude greater than the leading peak amplitude (cf. Fig. 22).

The coda for the transoceanic tsunami wave, taken as a whole, is not only developed from the dispersion as described above, but also the effects of reflection, particularly for continental subduction zone sources (Abe, 2000), scattering, and wave trapping. The foregoing analysis demonstrates that for constant water depth, the shape of the leading wave persists in the far field. Even for irregular water depth, observations indicate that the first cycle or so of the leading wave largely remains intact from the source region, such that the far-field wave has not "forgotten" its initial conditions (Carrier, 1970, 1971). Both circular (e.g. seamounts) and linear (e.g. escarpments) irregularities can cause scattering, depending on the water depth of the scattering feature relative to the basin water depth and the wavelength of the tsunami, thus leading to an increase in the duration of the tsunami coda as demonstrated analytically by Mofjeld et al. (2000b, 2001) and for the 2006 Kuril tsunami in particular by Koshimura et al. (2008) and Kowalik et al. (2008). In addition to scattering, wave trapping can occur along ridges and deflection along trenches, for the appropriate incidence angle (Chao, 1971; Mei, 1989), and focusing associated with spherically shaped irregularities may occur (Berry, 2007; Janssen et al., 2008), all of which can influence the coda of the tsunami.

Analytic expressions of tsunami scattering have taken both deterministic and statistical forms. A deterministic scattering index (S) is described by Mofjeld et al. (2000b, 2001) that is based on the minimum transmissivity for a ridge-like topographic feature:

$$S = 1 - \frac{2\varepsilon}{1+\varepsilon^2}, \quad \text{where } \varepsilon = \left(\frac{h_1}{h_0}\right)^{1/2}. \tag{42}$$

The advantage of this type of index is that it only depends on the depth to the top of the ridge h_1 relative to the depth of the ocean floor surrounding the ridge h_0. Therefore, a scattering index map can be developed based on the known bathymetry of an ocean (Mofjeld et al., 2001, 2004).

A statistical approach to tsunami scattering has been recently developed by Saito and Furumura (2009a). In this case, they treated bathymetric variations that cause scattering as a random medium (cf. Carrier (1970); Mysak (1978)), namely, that of a von Kármán autocorrelation function. The power spectral density of the bathymetry with radial wavenumber $\tilde{k}$ is given by

$$P\left(\tilde{k}\right) = \frac{4\pi\kappa\varepsilon^2 a^2}{\left(1+a^2\tilde{k}^2\right)^{\kappa+1}}, \tag{43}$$

where in this case ε is the rms value of the bathymetric variation and κ and a are the spectral decay exponent and the correlation distance, respectively. Note that this statistical model has also been used for the coseismic slip distribution on faults as described in Section 3.1. The corresponding scattering coefficient (g) derived by Saito and Furumura (2009a) as a function of azimuth about the scatterer (ϕ) and tsunami wavenumber (k) is

$$g(\phi, k) = \cos^2\phi \frac{\pi\kappa\varepsilon^2}{a} \frac{(ak)^3}{\left[1+a^2k^2\cos^2(\phi/2)\right]^{\kappa+1}}. \tag{44}$$

They indicate two end-member regimes for scattering, in terms of coda energy and attenuation coefficient: $ak \ll 1$ and $ak \gg 1$. For the first, more common, case in which the tsunami wavelength is much greater than the correlation distance, there is a symmetric radiation pattern (i.e. both back and forward scattering given by the $\cos^2\phi$ term; see Fig. 25a) and the attenuation coefficient is independent of the power-law exponent measure of seafloor roughness (κ). For $ak = 1$, forward scattering dominates (Fig. 25b) and the attenuation coefficient is dependent on κ. Janssen et al. (2008) discussed spatially coherent wave interference patterns associated with forward scattering and the effect on wave statistics.

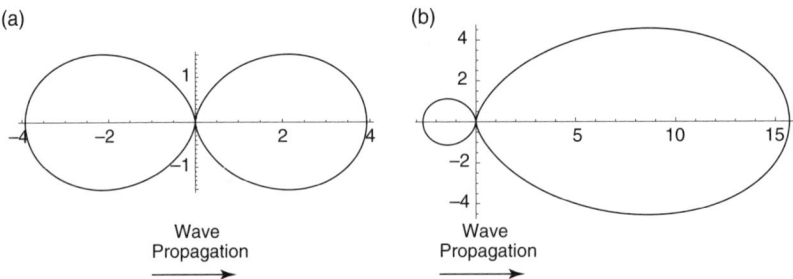

FIG. 25. Comparison of tsunami scattering radiation patterns derived by Saito and Furumura (2009a) for two cases: (a) $ak \ll 1$; (b) $ak = 1$.

4.2. Far-field Coastal Interaction

HYPOTHESIS 7. *Maximum amplitude and runup at the coast from a far-field tsunami are derived from the complex interaction of the long tsunami coda and the excitation and resonance of trapped edge waves and shelf modes.*

After the first arrival of the tsunami onshore, the shelf and coastal response greatly modifies the waveform characteristics from that observed in deep water. Coastal observations of far-field tsunamis are the result of a complex interaction between the coda developed during propagation, oblique propagation across the continental shelf, and the excitation of edge waves caused by coastline irregularities and the subsequent scattering and resonance. The result is a long coastal tsunami coda, with an *e*-folding time much greater than that for deep-ocean observations. I compare single-station records from tsunamis in the near-field oblique and far-field regimes and describe the observed effects based on analytic theory described in past studies.

A comparison of near-field oblique and far-field tsunami records at the Crescent City tide gage station yields information with which to evaluate Hypothesis 7. For this, I compare a near-field oblique record (1992 Cape Mendocino earthquake) (González et al., 1995) with far-field records (the 2006 Kuril and 1960 Chile earthquakes) (Fig. 26). The far-field records (Fig. 26b, c) are generally more resonant (i.e. less irregular than the near-field record), possibly due to prolonged edge wave excitation and resonance from the coda arriving at the coast. For the near-field oblique case, edge waves principally emanate from the Cape Mendocino earthquake to the south, whereas edge waves excited by the far-field tsunami possibly emanate to both the north and the south from secondary sources (i.e. coastline irregularities) (cf. Munk et al. (1964); Guza and Davis (1974); Fuller and Mysak (1977)). However, local resonance of trapped waves is particularly strong to the south of Crescent City (Horrillo et al., 2008), suggesting that the difference between the near-field oblique and far-field records is primarily caused by the longer open-ocean coda of the far-field tsunami.

Because of the oblique propagation path for the 1960 Chile tsunami, the record shown in Fig. 26c may have a large component of leaky (non-trapped) shelf modes, similar to what was observed for this tsunami in southern California (Miller et al., 1962; Snodgrass et al., 1962). The 2005 Crescent City tide gage record of a small tsunami generated by

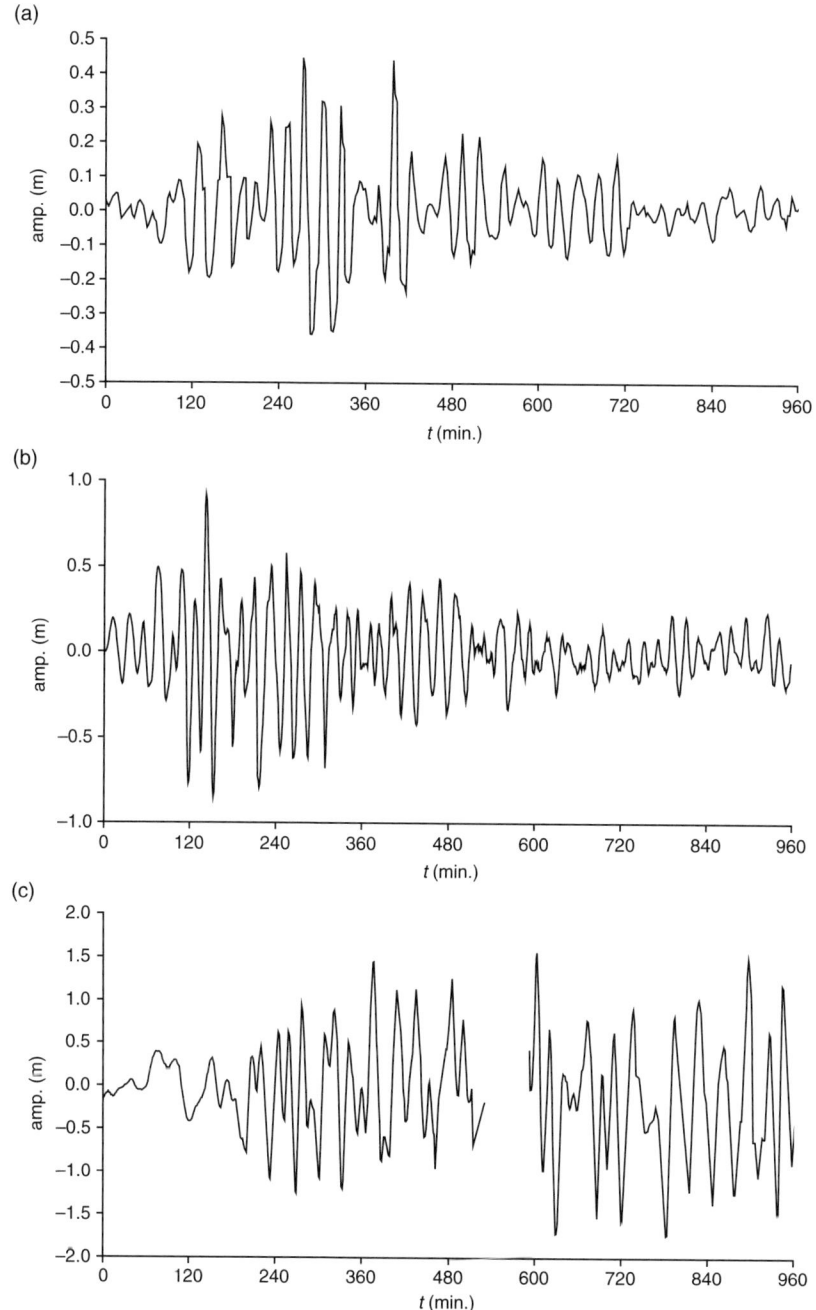

FIG. 26. Tide gage records at Crescent City for tsunamis generated by (a) a near-field oblique source, the 1992 Cape Mendocino earthquake and two far-field sources; (b) the 2006 Kuril earthquake and (c) the 1960 Chile earthquake.

a $M_w = 7.2$ strike-slip earthquake in the Gorda plate (approx. 3000 m water depth) is also explained by a larger component of leaky shelf waves compared to edge waves by Rabinovich et al. (2006). The strike of the 2005 Gorda plate earthquake is at a high angle to both the strike of the Cascadia inter-plate thrust (i.e. the source fault for the 1992 Cape Mendocino earthquake) and the trend of the shelf edge, suggesting predominantly oblique propagation. However the quadrupole tsunami radiation pattern for strike-slip earthquakes (González et al., 1991) complicates the analysis of trapped and shelf modes for this event.

As with the deep-ocean tsunami, a test can be performed to determine whether the coastal tsunami coda can be described as a time-varying Gaussian distribution (cf. Takahara and Yomogida (1992)). As before, I examine data from the 2006 Kuril Islands tsunami for both an island station (Midway Island) and a continental station (Crescent City) (Fig. 27). Estimation of the coda envelope indicates that the e-folding time for Midway Island is 22 hr, consistent with the general decay law with an e-folding time of 22.0 ± 0.7 hr measured by van Dorn (1984) for 28 tsunami sources and various station locations in the Pacific. The empirical density distribution of water-level elevations is shown in Fig. 28. According to the Shapiro-Wilk test, the Gaussian null hypothesis cannot be rejected for either the Crescent City or the Midway Island data. However, the empirical wave heights at Midway Island conform to a Rayleigh distribution better than wave heights at Crescent City (Fig. 29), suggesting that the narrow-band assumption used in the Longuet-Higgins (1952) derivation of the Rayleigh distribution may not hold for Crescent City.

The spectra for coastal stations (Fig. 30) show a strong site response that is also apparent in the spectra of tide gage records in response to background (non-tsunami) long waves (dashed lines). Rabinovich (1997) analyzed the spectra of three events along the Hokkaido-Kuril subduction zone and determined that the local hydrodynamic response is strongly controlled by the bathymetry near the station and is the dominant signal in the spectra (see also van Dorn (1984)). Rabinovich (1997) also separated the source and site response components by calculating the spectral ratios between background and tsunami spectra and found that the dominant period of the tsunami source is consistent with the earthquake dimensions and water depth above the source (i.e. taking into account Green's law discussed in Section 3.3). Rabinovich and Thomson (2007) analyzed 45 coastal records from the 2004 Indian Ocean tsunami and found that the site response for island stations is much less apparent than that for continental stations, owing to the absence of trapped shelf waves (see also Watanabe (1972); van Dorn (1987)). Moreover Rabinovich and Thomson (2007) noted that the first arrival is typically largest for broadside stations at regional distances (cf. Fig. 9), whereas the largest arrival occurs in the tsunami coda for all other stations, which is consistent with Hypotheses 1 and 4, respectively. They noted that the deep-ocean tsunami coda "pumps" energy into the local site response controlled by the nearby bathymetry, which is consistent with Hypothesis 7.

van Dorn (1984, 1987) suggested that the coastal response is related to normal-mode forcing across the continental shelf from the open-ocean tsunami coda (with an isotropic spectra, presumably caused by mid-ocean scattering), rather than the excitation of resonant edge waves as suggested above. However, in a detailed analysis of the coastal eigenmodes near Crescent City, Horrillo et al. (2008) indicated that both impedance matching of the gravest shelf mode and transient edge wave modes (cf. Fuller and Mysak

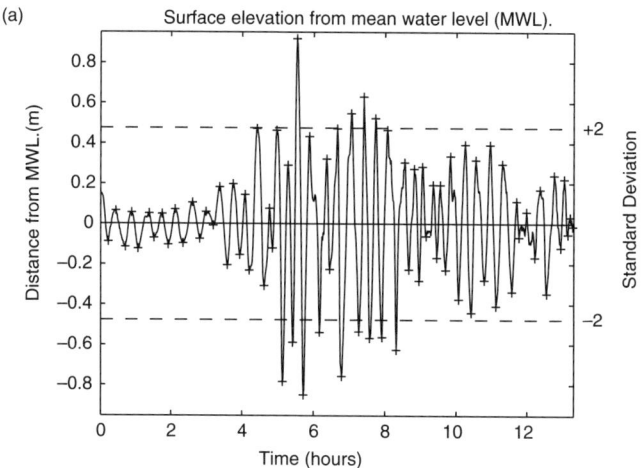

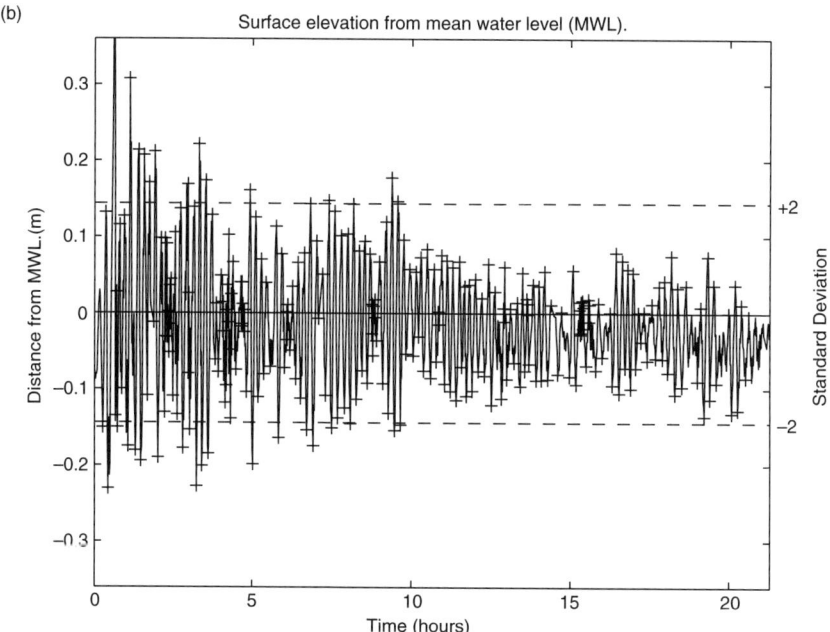

FIG. 27. Time series of tide gage measurements of the 15 November 2006 Kuril tsunami from locations indicated in Fig. 16: (a) Crescent City; (b) Midway Island. Location of crest and trough amplitudes indicated by +'s.

(1977)) contribute to the complex tide gage record. Horrillo *et al.* (2008) indicated that resonance within the Crescent City harbor does not appear to be a major factor, owing to the fact that the observed dominant periods are greater than the natural period of the

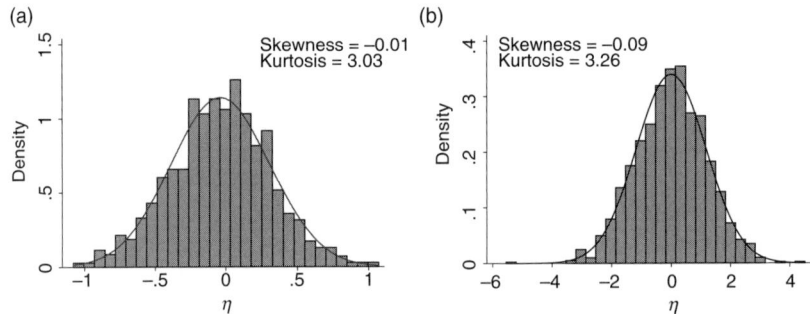

FIG. 28. Empirical density distribution of tsunami coda waves from data plotted in Fig. 27: (a) Crescent City; (b) Midway Island. Line indicates the Gaussian distribution model.

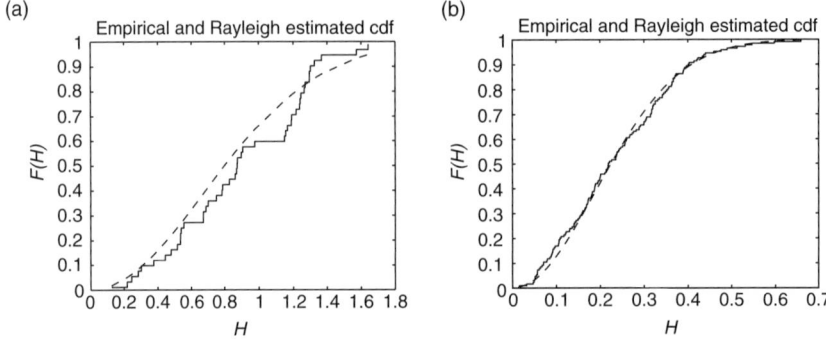

FIG. 29. Empirical cumulative distribution (solid line) of wave height (H) from the data shown in Fig. 27: (a) Crescent City; (b) Midway Island. Line indicates the Rayleigh distribution model.

harbor. (It should be noted that harbor resonance is an important component of the local site response for many other tide gage locations.) Shelf resonance has also been used to explain the persistent signal (over 2 days) in Brazil from the 2004 tsunami generated by the Sumatra-Andaman earthquake (França and de Mesquita, 2007).

Whereas the origin of edge wave development from a near-field source at the coast can be analyzed from theoretical studies (Kajiura, 1972; Ishii and Abe, 1980; Fujima et al., 2000), it is less clear how edge waves are developed for far-field tsunamis. For a uniform coast line, distant tsunami waves will specularly reflect without the excitation of edge waves (i.e. Case 3 of Fujima et al. (2000)). For an irregular coastline, however, it has been proposed that edge waves can be excited by different mechanisms. Guza and Davis (1974) showed that nonlinear coupling of a triad composed of the incident wave and two oppositely propagating edge waves can result in an exponential growth rate of edge waves at specific frequencies. Fuller and Mysak (1977) suggested that resonant interaction of the incident wave with the wavenumber spectrum of coastal irregularities can also excite edge waves. As with edge waves developed in the near-field oblique regime, it is difficult

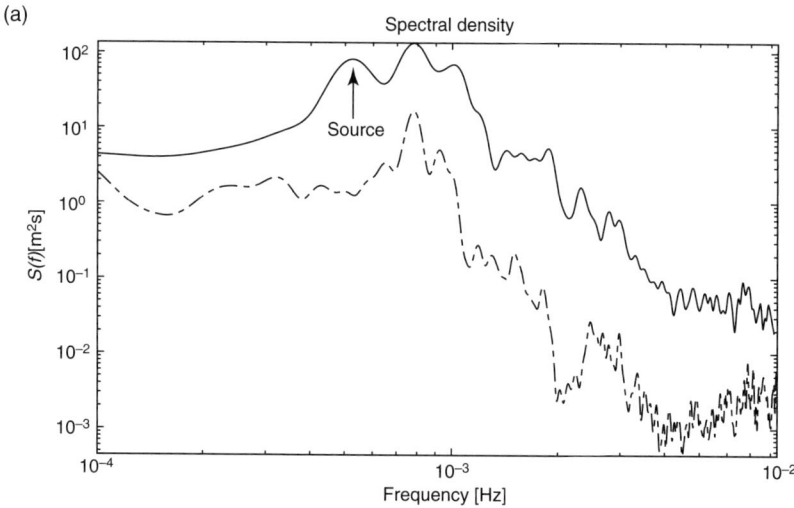

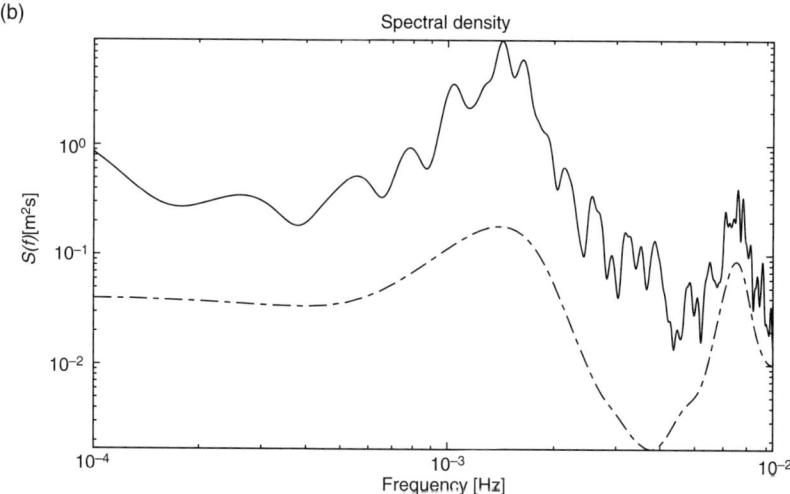

FIG. 30. Spectra of tsunami coda waves from the data plotted in Fig. 27: (a) Crescent City; (b) Midway Island. Background wave spectra are shown by a dashed line.

to trace the origin of individual phases observed on tide gage records, particularly when the incident wave has a long coda.

Far-field maximum amplitudes at the coast are particularly difficult to predict because of the interaction of the tsunami coda with the coastal response. As a notable example, Watanabe (1972) examined the maximum far-field amplitude associated with four major trans-Pacific tsunamis: those generated by 1952 Kamchatka, 1957 Aleutian, 1960 Chile, and 1964 Gulf of Alaska earthquakes. Contrary to a fundamental understanding of attenuation during tsunami propagation (let alone the exact decay exponent provided by Eqs (31) and (38)), Watanabe (1972) indicated that the *maximum* far-field amplitude observed at coastal stations *increases* with the travel time to those stations for the 1952, 1960, and 1964 tsunamis. While one could argue that the scatter in the data does not present statistically significant results, the point remains that a simple understanding of tsunami physics in the deep ocean, such as attenuation with travel time, does not translate to similar predictions for maximum tsunami amplitudes at the coast that are associated with late arrivals from, for example, multiple trapped waves and are related to a site-specific response.

5. Summary and Discussion

5.1. Observations and Hypotheses

In this paper, past observations of tsunami events have been described according to a broad understanding of tsunami physics. For tsunami generation, the focus has been on earthquakes, owing to the frequency of occurrence of seismogenic tsunami sources and to a lack of data on tsunami generation from other sources, such as submarine landslides. The primary observation related to tsunami generation from earthquakes is spatially heterogeneous coseismic slip that is determined from the inversion of seismic waveforms. Observations relating to tsunami propagation and runup, are divided among three coastal regimes: near-field broadside, near-field oblique, and far field (Fig. 1). This division is based on the distinctly different tsunami physics that control the amplitude and timing of the maximum observed wave (Carrier, 1995).

Several working hypotheses have been presented in this paper to describe the observations, primarily relating to the amplitude (or runup) and timing of the largest tsunami wave, which are of primary interest in developing tsunami hazard assessments (e.g. see González *et al.* (in press)). In the near-field broadside regime and for a nominally regular coast, the maximum amplitude of a tsunami is most often associated with the direct arrival from the source. Accordingly, spatial heterogeneity of the source appears to have a significant effect on the spatial distribution of broadside runup. Moreover, the severity of the tsunami is dependent on the water depth overlying regions of high slip on the fault. For high slip beneath deep water, more shoaling amplification and higher tsunami runups result, as indicated, for example, by a comparison of the tsunamis generated by the 1952 and 2003 Tokachi-Oki earthquakes (Fig. 12).

In the near-field oblique regime, the maximum amplitude of a tsunami is most often associated with excited edge waves that propagate at significantly slower phase and group speeds than non-trapped modes. The details of how near-field edge waves are excited depend on the proximity of the initial wavefield to the shoreline and the orientation of initial wave crest/trough relative to the shoreline (Fujima *et al.*, 2000). Typically, the orientation of the waves for seismogenic tsunamis is subparallel to the shoreline, exciting predominantly fundamental mode edge waves, whereas for landslide tsunamis, the orientation is subnormal to the shoreline, exciting higher edge wave modes.

Finally, a tsunami coda gradually develops in the far field as a result of dispersion and scattering from bathymetric features. As the tsunami arrives at transoceanic coasts, energy from the coda is pumped into the nearshore response that includes the excitation of edge waves, scattering from coastline irregularities, and resonance of trapped modes (e.g. see Rabinovich and Thomson (2007)). In this case, the tsunami is very persistent in time and the maximum amplitude can occur much later than the first arrival.

5.2. Statistical Descriptions

Rather than treating the evolution of tsunamis from a strictly deterministic approach, I have reviewed different statistical models that have been developed from generation through runup. Unexpected behavior of tsunamis is understood better from a statistical perspective, compared to simple deterministic models of tsunami physics (e.g. see Fig. 2). For coseismic slip, a 2D stochastic model initially developed by Andrews (1980) has been extensively used in describing slip distributions derived from seismic waveform inversions. Although in theory, statistical descriptions of runup observations can be linked to the stochastic slip model, the sparse and non-uniform nature of runup observations from post-tsunami surveys, as well as the complex nature of nearshore wave dynamics, presently preclude such a link. The log-normal model proposed by Choi *et al.* (2002, 2006) is currently the best constrained statistical description of runup distribution. The log-normal distribution (and the power-law distribution in select cases) indicates a higher likelihood of extreme runup over what would be expected from a Gaussian distribution of runup tied to the potency of the source (e.g. seismic moment).

In the far field, the observed tsunami coda can be described by an exponentially decaying Gaussian distribution of wave elevations, similar to seismic coda statistical models (e.g. see Takahara and Yomogida (1992)). For narrow-banded waves, the distribution of wave amplitudes therefore corresponds to a Rayleigh distribution. Such a model encompasses frequent observations of a delay in the arrival of the maximum amplitude phase after the first (direct) arrival. The exponential decay time is greater for nearshore observations compared to deep-ocean measurements, owing to the excitation, scattering, and resonance of edge waves and leaky shelf modes as part of the overall nearshore site response.

The statistical models used to describe tsunamis are intended to encompass possible outcomes that result from either highly complex physics (i.e. aleatory uncertainty) or for linear aspects of tsunami physics where insufficient data exist for accurately developing deterministic solutions (i.e. epistemic uncertainty). For the former, the rate- and state-friction law for fault rupture is sufficiently complex for generating multiple slip realizations for only slight differences in initial conditions, even for simple fault geometries. Real fault systems are likely to be geometrically complex (e.g. see Power and Tullis (1991)), with spatially heterogeneous pre-stress conditions and pore pressure distributions. Earthquakes with the same hypocenter and the same overall seismic moment are likely to have different slip distributions, as exemplified by the 1952 and 2003 Tokachi-Oki earthquakes (Fig. 12). Therefore, the stochastic coseismic slip model is intended to provide an ensemble of realizations representing aleatory uncertainty in understanding earthquake rupture dynamics from a probabilistic perspective. In contrast, the difficulty in accurately simulating the tsunami coda may be somewhat surprising, given that the physics of long-wave propagation is fairly well understood. It is likely that much of the coda arises from scattering in the open ocean and edge wave excitation along coasts, for which detailed bathymetry is needed. In this case, statistical models of the far-field waveform encompass epistemic uncertainty related to the data needed to accurately simulate nearshore wave physics.

5.3. Knowledge Gaps

Our understanding of tsunami physics is far from complete, owing in part to obvious gaps in observing certain phases of tsunami evolution. With regard to tsunami generation, there are very few direct measurements of submarine landslide movement. This information is critical for estimating the constitutive parameters for landslide dynamics and the efficiency at which landslides generate tsunamis. For seismogenic tsunamis, there are also few direct measurements of coseismic vertical displacement of the seafloor, although the unique measurements stemming from the 2003 Tokachi-Oki earthquake (Mikada *et al.*, 2006), as well as many onshore case studies, suggest that elastic displacement models driven by heterogeneous coseismic slip on the fault plane are adequate. With regard to tsunami propagation, we do not have a dense spatial observation network of deep-ocean tsunami wave heights, although the network of BPR stations throughout the world's oceans is greatly expanding (Mofjeld, 2009). In addition, space-borne technology including satellite altimetry and detection of "tsunami shadows" can greatly help fill this gap (Godin, 2004; Godin *et al.*, 2009). When the tsunami arrives at the coast, we also have very little information regarding the time evolution of the runup process, although we are gaining a better understanding of this process in the field through observations of tsunami deposits combined with advanced sediment transport models (Apotsos *et al.*, 2009). There are also gaps related to a theoretical understanding of tsunami generation physics. In solid-earth physics, the outstanding question relates to the fault mechanics of tsunami earthquakes and splay fault ruptures, particularly with regard to what conditions (friction parameters, pore pressure, pre-stress conditions, etc.) are needed for these types of earthquakes to occur at very shallow depths (cf. Taylor (1998); Kame *et al.* (2003)). As data from new tsunami observation technologies are obtained and as we gain a better theoretical understanding of tsunami generation and hydrodynamic processes, scientists will have a broader perspective of what to expect when a tsunami occurs.

ACKNOWLEDGMENTS

The author gratefully acknowledges careful reading of this paper and constructive reviews by Frank González, Alex Apotsos, and the Advances in Geophysics editor, Renata Dmowska. This study could not have been performed without access to databases of tsunami observations, including the NOAA event archive of DART bottom pressure recorder data, digital tide gage records at the West Coast/Alaska Tsunami Warning Center, and analog tide gage records from the Earthquake Research Institute at the University of Tokyo. Analysis of tsunami records was performed in part using the Wave Analysis for Fatigue and Oceanography (WAFO) package developed at Lund University, Sweden (The WAFO Group, 2000). Map figures were created using Generic Mapping Tools (GMT) (Wessel and Smith, 1995).

References

Abe, K., Ishii, H. (1987). Distribution of maximum water levels due to the Japan Sea tsunami on 26 May 1983. *J. Oceanogr. Soc. Japan* **43**, 169–182.
Abe, K., Abe, K., Tsuji, Y., Imamura, F., Katao, H., Iio, Y., Satake, K., Bourgeois, J., Noguera, E., Estrada, F. (1993). Field survey of the Nicaragua earthquake and tsunami of September 2, 1992. *Bull. Earthquake Res. Inst.* **68**, 23–70.
Abe, K. (2000). Predominance of long periods in large Pacific tsunamis. *Sci. Tsunami Hazards* **18**, 15–34.
Abercrombie, R.E., Antolik, M., Flezer, K., Ekström, G. (2001). The 1994 Java tsunami earthquake: Slip over a subducting seamount. *J. Geophys. Res.* **106**, 6595–6607.
Aki, K., Chouet, B. (1975). Origin of coda waves: Source, attenuation, and scattering effects. *J. Geophys. Res.* **80**, 3322–3342.
Aki, K., Richards, P.G. (1980). Quantitative Seismology: Theory and Methods. W. H. Freeman and Company, San Francisco, 557 pp.
Ammon, C.J., Ji, C., Thio, H.K., Robinson, D., Ni, S., Hjorleifsdottir, V., Kanamori, H., Lay, T., Das, S., Helmberger, D., Ichinose, G., Polet, J., Wald, D. (2005). Rupture process of the 2004 Sumatra–Andaman earthquake. *Science* **308**, 1133–1139.
Ammon, C.J., Kanamori, H., Lay, T., Velasco, A.A. (2006). The 17 July 2006 Java tsunami earthquake. *Geophys. Res. Lett.* **33**, doi:10.1029/2006GL028005.
Andrews, D.J. (1980). A stochastic fault model 1. Static case. *J. Geophys. Res.* **85**, 3867–3877.
Apotsos, A., Jaffe, B., Gelfenbaum, G., Elias, E. (2009). Modeling time-varying tsunami sediment deposition, paper presented at Coastal Dynamics 2009: Impacts of Human Activities on Dynamic Coastal Processes, Tokyo, Japan, doi:10.1142/9789814282475_0037.
Archuleta, R.J. (1984). A faulting model for the 1979 Imperial Valley earthquake. *J. Geophys. Res.* **89**, 4559–4585.
Baba, T., Cummins, P.R., Hori, T. (2005). Compound fault rupture during the 2004 off the Kii Peninsula earthquake (M 7.4) inferred from highly resolved coseismic sea-surface deformation. *Earth Planets Space* **57**, 167–172.
Baba, T., Cummins, P.R., Hori, T., Kaneda, Y. (2006a). High precision slip distribution of the 1944 Tonankai earthquake inferred from tsunami waveforms: Possible slip on a splay fault. *Tectonophys.* **426**, 119–134.
Baba, T., Hirata, K., Hori, T., Sakaguchi, H. (2006b). Offshore geodetic data conducive to the estimation of the afterslip distribution following the 2003 Tokachi–Oki earthquake. *Earth Planet. Sci. Lett.* **241**, 281–292.
Banerjee, P., Pollitz, F.F., Nagarajan, B., Bürgmann, R. (2007). Coseismic slip distributions of the 26 December 2004 Sumatra–Andaman and 28 March 2005 Nias earthquake from GPS static offsets. *Bull. Seismol. Soc. Am.* **97(1A)**, S86–S102.
Baptista, A.M., Priest, G.R., Murty, T.S. (1993). Field survey of the 1992 Nicaragua tsunami. *Marine Geodesy* **16**, 169–203.
Beeler, N.M., Tullis, T.E., Weeks, J.D. (1994). The roles of time and displacement in the evolution effect in rock friction. *Geophys. Res. Lett.* **21**, 1987–1990.
Ben-Menahem, A., Rosenman, M. (1972). Amplitude patterns of tsunami waves from submarine earthquakes. *J. Geophys. Res.* **77**, 3097–3128.
Ben-Zion, Y., Rice, J.R. (1997). Dynamic simulations of slip on a smooth fault in an elastic solid. *J. Geophys. Res.* **102**, 17,771–717,784.
Bent, A.L. (1995). A complex double-couple source mechanism for the Ms 7.2 1929 Grand Banks earthquake. *Bull. Seismol. Soc. Am.* **85**, 1003–1020.
Beresnev, I.A. (2003). Uncertainties in finite-fault slip inversions: To what extent to believe? (A critical review). *Bull. Seismol. Soc. Am.* **93**, 2445–2458.
Berge, C., Gariel, J.C., Bernard, P. (1998). A very broad-band stochastic source model used for near source strong motion prediction. *Geophys. Res. Lett.* **25**, 1063–1066.
Berry, M.V. (2007). Focused tsunami waves. *Proc. R. Soc. Lond. A* **463**, 3055–3071.
Bhat, H.S., Olives, M., Dmowska, R., Rice, J.R. (2007). Role of fault branches in earthquake rupture dynamics. *J. Geophys. Res.* **112**, doi:10.1029/2007JB005027.

Bilek, S.L., Lay, T. (1999). Rigidity variations with depth along interplate megathrust faults in subduction zones. *Nature* **400**, 443–446.
Bilek, S.L., Lay, T. (2000). Depth dependent rupture properties in circum-Pacific subduction zones. In: Rundle, J.B., et al. (Eds.), GeoComplexity and the Physics of Earthquakes. American Geophysical Union, Washington, D.C, pp. 165–186.
Bilek, S.L., Lay, T. (2002). Tsunami earthquakes possibly widespread manifestations of frictional conditional stability. *Geophys. Res. Lett.* **29**, doi:10.1029/2002GL015215.
Bilek, S.L. (2007). Using earthquake rupture variations along the Sumatra–Andaman subduction system to examine fault zone variations. *Bull. Seismol. Soc. Am.* **97**, S62–S70.
Bird, P., Kagan, Y.Y. (2004). Plate-tectonic analysis of shallow seismicity: Apparent boundary width, beta-value, corner magnitude, coupled lithosphere thickness, and coupling in 7 tectonic settings. *Bull. Seismol. Soc. Am.* **94**, 2380–2399.
Borrero, J.C. (2001). Changing field data gives better model results: An example from Papua New Guinea, paper presented at International Tsunami Symposium 2001, Seattle, Washington, pp. 397–405.
Brodsky, E.E., Gordeev, E., Kanamori, H. (2003). Landslide basal friction as measured by seismic waves. *Geophys. Res. Lett.* **30**, doi:10.1029/2003GL018485.
Carrier, G.F., Greenspan, H.P. (1958). Water waves of finite amplitude on a sloping beach. *J. Fluid Mech.* **4**, 97–109.
Carrier, G.F. (1970). Stochastically driven dynamical systems. *J. Fluid Mech.* **44**, 249–264.
Carrier, G.F. (1971). The dynamics of tsunamis. In: Reid, W.H. (Ed.), Mathematical Problems in the Geophysical Sciences. American Mathematical Society, Providence, Rhode Island, pp. 157–187.
Carrier, G.F. (1995). On-shelf tsunami generation and coastal propagation. In: Tsuchiya, Y., Shuto, N. (Eds.), Tsunami: Progress in Prediction, Disaster Prevention and Warning. Kluwer, Dordrecht, The Netherlands, pp. 1–20.
Carrier, G.F., Wu, T.T., Yeh, H. (2003). Tsunami run-up and draw-down on a plane beach. *J. Fluid Mech.* **475**, 79–99.
Cattin, R., Briole, P., Lyon-Caen, H., Bernard, P., Pinettes, P. (1999). Effects of superficial layers on coseismic displacements for a dip-slip fault and geophysical implications. *Geophys. J. Int.* **137**, 149–158.
Chao, Y.Y. (1971). An asymptotic evaluation of the wave field near a smooth caustic. *J. Geophys. Res.* **76**, 7401–7408.
Chen, Q., Kirby, J.T., Dalrymple, R.A., Kennedy, A.B., Chawla, A. (2000). Boussinesq modeling of wave transformation, breaking, and runup: II. 2D. *Journal of Waterway, Port, Coastal, and Ocean Engineering* **126**, 48–56.
Chlieh, M., Avouac, J.P., Hjorleifsdottir, V., Song, T.A., Ji, C., Sieh, K., Sladen, A., Hebert, H., Prawirodirdjo, L., Bock, Y., Galetzka, J. (2007). Coseismic slip and afterslip of the great (Mw 9.15) Sumatra–Andaman earthquake of 2004. *Bull. Seismol. Soc. Am.* **97(1A)**, S152–S173.
Choi, B.H., Pelinovsky, E., Ryabov, I., Hong, S.J. (2002). Distribution functions of tsunami wave heights. *Nat. Haz.* **25**, 1–21.
Choi, B.H., Hong, S.J., Pelinovsky, E. (2006). Distribution of runup heights of the December 26, 2004 tsunami in the Indian Ocean. *Geophys. Res. Lett.* **33**, doi:10.1029/2006GL025867.
Coussot, P., Meunier, M. (1996). Recognition, classification and mechanical description of debris flows. *Earth-Sci. Rev.* **40**, 209–227.
Coussot, P. (1997). Mudflow Rheology and Dynamics. A. A. Balkema, Rotterdam, 255 pp.
Coussot, P., Laigle, D., Arattano, M., Deganutti, A., Marchi, L. (1998). Direct determination of rheological characteristics of debris flow. *J. Hydraulic Eng.* **124**, 865–868.
Cummins, P.R., Kaneda, Y. (2000). Possible splay fault slip during the 1946 Nankai earthquake. *Geophys. Res. Lett.* **27** (17), 2725–2728.
Dahlen, F.A. (1993). Single-force representation of shallow landslide sources. *Bull. Seismol. Soc. Am.* **83**, 130–143.
Dahlen, F.A., Tromp, J. (1998). Theoretical Global Seismology. Princeton University Press, Princeton, New Jersey, 1025 pp.
Dengler, L., Uslu, B., Barberopoulou, A., Yim, S.C., Kelly, A. (2009). The November 15, 2006 Kuril Islands-generated tsunami in Crescent City, California. *Pure Appl. Geophys.* **166**, 37–53.

Dieterich, J.H. (1979). Modeling of rock friction. Part 1. Experimental results and constitutive equations. *J. Geophys. Res.* **84**, 2161–2168.
Dieterich, J.H. (1995). Earthquake simulations with time-dependent nucleation and long-range interactions. *Nonlinear Process. Geophys.* **2**, 109–120.
Dieterich, J.H. (2007). Applications of rate- and state-dependent friction to models of fault slip and earthquake occurrence. In: Kanamori, H. (Ed.), Treatise on Geophysics. Elsevier, pp. 107–129.
Dmowska, R., Kostrov, B.V. (1973). A shearing crack in a semi-space under plane strain conditions. *Arch. Mech.* **25**, 421–440.
Dmowska, R., Rice, J.R. (1986). Fracture theory and its seismological applications. In: Teisseyre, R. (Ed.), Continuum Theories in Solid Earth Physics. PWN-Polish Scientific Publishers, Warsaw, pp. 187–255.
Dmowska, R., Rice, J.R., Lovison, L.C., Josell, D. (1988). Stress transfer and seismic phenomena in coupled subduction zones during the earthquake cycle. *J. Geophys. Res.* **93**, 7869–7884.
Elverhøi, A., Issler, D., De Blasio, F.V., Ilstad, T., Harbitz, C.B., Gauer, P. (2005). Emerging insights into the dynamics of submarine debris flows. *Natural Hazards and Earth System Sciences* **5**, 633–648.
Engdahl, E.R., Dewey, J.W., Fujita, K. (1982). Earthquake location in island arcs. *Phys. Earth Planet. Int.* **30**, 145–156.
Farreras, S.F. (2000). Post-tsunami field survey procedures: An outline. *Nat. Haz.* **21**, 207–214.
Fernández-Nieto, E.D., Bouchut, F., Bresch, D., Castro, M.J., Diaz, A., Mangeney (2008). A new Savage-Hutter type model for submarine avalanches and generated tsunami. *J. Comput. Phys.* **227**, 7720–7754.
Fine, I.V., Rabinovich, A.B., Bornhold, B.D., Thomson, R., Kulikov, E.A. (2005). The Grand Banks landslide-generated tsunami of November 18, 1929: preliminary analysis and numerical modeling. *Mar. Geol.* **215**, 45–57.
Flück, P., Hyndman, R.D., Wang, K. (1997). Three-dimensional dislocation model for great earthquakes of the Cascadia subduction zone. *J. Geophys. Res.* **102**, 20,539–20,550.
França, C.A.S., de Mesquita, A.R. (2007). The December 26th 2004 tsunami recorded along the southeastern coast of Brazil. *Nat. Haz.* **40**, 209–222.
Frankel, A.D. (1991). High-frequency spectral falloff of earthquakes, fractal dimension of complex rupture, b value, and the scaling of strength on faults. *J. Geophys. Res.* **96**, 6291–6302.
Freund, L.B., Barnett, D.M. (1976). A two-dimensional analysis of surface deformation due to dip-slip faulting. *Bull. Seismol. Soc. Am.* **66**, 667–675.
Fujii, Y., Satake, K. (2007). Tsunami source of the 2004 Sumatra–Andaman earthquake inferred from tide gauge and satellite data. *Bull. Seismol. Soc. Am.* **97**, S192–S207.
Fujii, Y., Satake, K. (2008). Tsunami sources of the November 2006 and January 2007 Great Kuril earthquakes. *Bull. Seismol. Soc. Am.* **98**, 1559–1571.
Fujima, K., Dozono, R., Shigemura, T. (2000). Generation and propagation of tsunami accompanying edge waves on a uniform shelf. *Coastal Eng. J.* **42**, 211–236.
Fukao, Y. (1979). Tsunami earthquakes and subduction processes near deep-sea trenches. *J. Geophys. Res.* **84**, 2303–2314.
Fukuyama, E. (2009). Fault-zone Properties and Earthquake Rupture Dynamics. Elsevier, Inc, London, 336 pp.
Fuller, J.D., Mysak, L.A. (1977). Edge waves in the presence of an irregular coastline. *J. Phys. Oceanogr.* **7**, 846–855.
Geist, E.L. (1999). Local tsunamis and earthquake source parameters. *Adv. Geophys.* **39**, 117–209.
Geist, E.L., Dmowska, R. (1999). Local tsunamis and distributed slip at the source. *Pure Appl. Geophys.* **154**, 485–512.
Geist, E.L., Bilek, S.L. (2001). Effect of depth-dependent shear modulus on tsunami generation along subduction zones. *Geophys. Res. Lett.* **28** (7), 1315–1318.
Geist, E.L. (2002). Complex earthquake rupture and local tsunamis. *J. Geophys. Res.* **107**, doi:10.1029/2000JB000139.
Geist, E.L., Yoshioka, S. (2004). Effect of structural heterogeneity and slip distribution on coseismic vertical displacement from rupture on the seattle fault. Open-File Report 2004-1010, U.S. Geological Survey.
Geist, E.L., Bilek, S.L., Arcas, D., Titov, V.V. (2006). Differences in tsunami generation between the December 26, 2004 and March 28, 2005 Sumatra earthquakes. *Earth Planets Space* **58**, 185–193.

Geist, E.L., Titov, V.V., Arcas, D., Pollitz, F.F., Bilek, S.L. (2007). Implications of the December 26, 2004 Sumatra–Andaman earthquake on tsunami forecast and assessment models for great subduction zone earthquakes. *Bull. Seismol. Soc. Am.* **97**, S249–S270.

Geist, E.L., Lynett, P.J., Xu, J.P. (2009a). Re-examination of the Monterey Bay tsunami triggered by the 1989 Loma Prieta earthquake [abs.]. *Seismol. Res. Lett.* **80**, 334.

Geist, E.L., Parsons, T., ten Brink, U.S., Lee, H.J. (2009b). Tsunami Probability. In: Bernard, E.N., Robinson, A.R. (Eds.), In: The Sea, Vol. 15, Harvard University Press, Cambridge, Massachusetts, pp. 93–135.

Geller, R.J. (1976). Scaling relations for earthquake source parameters and magnitudes. *Bull. Seismol. Soc. Am.* **66**, 1501–1523.

Gica, E., Teng, M.H., Liu, P.L.-F., Titov, V.V., Zhou, H. (2007). Sensitivity analysis of source parameters for earthquake-generated distant tsunamis. *Journal of the Waterways and Harbors Division, A.S.C.E.* **133**, 429–441.

Godin, O.A. (2004). Air-sea interaction and feasibility of tsunami detection in the open ocean. *J. Geophys. Res.* **109**, doi:10.1029/2003JC002030.

Godin, O.A., Irisov, V.G., Leben, R.R., Hamlington, B.D., Wick, G.A. (2009). Variations in sea surface roughness induced by the 2004 Sumatra–Andaman tsunami. *Natural Hazards and Earth System Sciences* **9**, 1135–1147.

Golovachev, E.V., Kochergin, I.E., Pelinovsky, E.N. (1992). The effect of the Airy phase during propagation of edge waves. *Soviet J. Phys. Oceanogr.* **3**, 1–7.

González, F.I., Mader, C.L., Eble, M.C., Bernard, E.N. (1991). The 1987-88 Alaskan Bight tsunamis: Deep ocean data and model comparisons. *Nat. Haz.* **4**, 119–139.

González, F.I., Kulikov, Y.A. (1993). Tsunami dispersion observed in the deep ocean. In: Tinti, S. (Ed.), Tsunamis in the World. Kluwer Academic Publishers, Dordrecht, The Netherlands, pp. 7–16.

González, F.I., Satake, K., Boss, E.F., Mofjeld, H.O. (1995). Edge wave and non-trapped modes of the 25 April 1992 Cape Mendocino tsunami. *Pure Appl. Geophys.* **144**, 409–426.

González, F.I., Geist, E.L., Jaffe, B.E., Kânoglu, U., Mofjeld, H.O., Synolakis, C.E., Titov, V.V., Arcas, D., Bellomo, D., Carlton, D., Horning, T., Johnson, J., Newman, J.C., Parsons, T., Peters, R., Peterson, C., Priest, G.R., Venturato, A.J., Weber, J., Wong, F., Yalciner, A.C. Probabilistic tsunami hazard assessment at Seaside, Oregon for near- and far-field seismic sources, *J. Geophys. Res.* (in press).

Greenberg, D.A., Dupont, D., Lyard, F.H., Lynch, D.R., Werner, F.E. (2007). Resolution issues in numerical models of ocean and coastal circulation. *Cont. Shelf Res.* **27**, 1317–1343.

Guatteri, M., Mai, P.M., Beroza, G.C. (2004). A pseudo-dynamic approximation to dynamic rupture models for strong ground motion prediction. *Bull. Seismol. Soc. Am.* **94**, 2051–2063.

Guza, R.T., Davis, R.E. (1974). Excitation of edge waves by waves incident on a beach. *J. Geophys. Res.* **79**, 1285–1129.

Hammack, J.L. (1973). A note on tsunamis: Their generation and propagation in an ocean of uniform depth. *J. Fluid Mech.* **60**, 769–799.

Hanks, T.C. (1979). b values and $\omega^{-\gamma}$ seismic source models: Implications for tectonic stress variations along active crustal fault zones and the estimation of high-frequency strong ground motion. *J. Geophys. Res.* **84**, 2235–2242.

Hartzell, S.H., Heaton, T.H. (1985). Teleseismic time functions for large, shallow, subduction zone earthquakes. *Bull. Seismol. Soc. Am.* **75**, 965–1004.

Hashimoto, M., Choosakul, N., Hashizume, M., Takemoto, S., Takiguchi, H., Fukuda, Y., Fujimore, K. (2006). Crustal deformations associated with the great Sumatra–Andaman earthquake deduced from continuous GPS observation. *Earth Planets Space* **58**, 127–139.

Haskell, N.A. (1964). Total energy and energy spectral density of elastic wave radiation from propagating faults. *Bull. Seismol. Soc. Am.* **54**, 1811–1841.

Heki, K., Tamura, Y. (1997). Short term afterslip in the 1994 Sanriku–Haruka-Oki earthquake. *Geophys. Res. Lett.* **24**, 3285–3288.

Herrero, A., Bernard, P. (1994). A kinematic self-similar rupture process for earthquakes. *Bull. Seismol. Soc. Am.* **84**, 1216–1228.

Hirata, K., Aoyagi, M., Mikada, H., Kawaguchi, K., Kaiho, Y., Iwase, R., Morita, S., Fujisawa, I., Sugioka, H., Mitsuzawa, K., Suyehiro, K., Kinoshita, H. (2002). Real-time geophysical measurements on the deep seafloor using submarine cable in the southern Kuril subduction zone. *IEEE J. Ocean. Eng.* **27**, 170–181.

Hirata, K., Geist, E.L., Satake, K., Tanioka, Y., Yamaki, S. (2003). Slip distribution of the 1952 Tokachi–Oki earthquake (M 8.1) along the Kuril Trench deduced from tsunami waveform inversion. *J. Geophys. Res.* **108**, doi:10.1029/2002JB001976.
Hirata, K., Tanioka, Y., Satake, K., Yamaki, S., Geist, E.L. (2004). The tsunami source area of the 2003 Tokachi–Oki earthquake estimated from tsunami travel times and its relationship to the 1952 Tokachi–Oki earthquake. *Earth Planets Space* **56**, 367–372.
Hirata, K., Satake, K., Tanioka, Y., Kuragano, T., Hasegawa, Y., Hayashi, Y., Hamada, M. (2006). The 2004 Indian Ocean tsunami: Tsunami source model from satellite altimetry. *Earth Planets Space* **58**, 195–201.
Hisada, Y. (2000). A theoretical omega-square model considering the spatial variation in slip and rupture velocity. *Bull. Seismol. Soc. Am.* **90**, 387–400.
Hisada, Y. (2001). A theoretical omega-square model considering the spatial variation in slip and rupture velocity. Part 2: Case for a two-dimensional source model. *Bull. Seismol. Soc. Am.* **91**, 651–666.
Hogben, N. (1990). Long term wave statistics. In: Le Méhauté, B., Hanes, D.M. (Eds.), Ocean Engineering Science. John Wiley & Sons, New York, pp. 293–333.
Horrillo, J., Knight, W., Kowalik, Z. (2008). Kuril Islands tsunami of November 2006: 2. Impact at Crescent City by local enhancement. *J. Geophys. Res.* **113**, doi:10.1029/2007JC004404.
Huang, N.E., Tung, C.C., Long, S.R. (1990). The probability structure of the ocean surface. In: Le Méhauté, B., Hanes, D.M. (Eds.), Ocean Engineering Science. John Wiley & Sons, New York, pp. 335–366.
Hughes Clarke, J.E. (1990). Late stage slope failure in the wake of the 1929 Grand Banks earthquake. *Geo-Marine Letters* **10**, 69–79.
Hungr, O., Evans, S.G., Bovis, M.J., Hutchinson, J.F. (2001). A review of the classification of landslides of the flow type. *Environ. Eng. Geosci.* **7**, 221–238.
Ichinose, G., Somerville, P., Thio, H.K., Graves, R., O'Connell, D. (2007). Rupture process of the 1964 Prince William Sound, Alaska, earthquake from the combined inversion of seismic, tsunami, and geodetic data. *J. Geophys. Res.* **112**, doi:10.1029/2006JB004728.
Ihmlé, P.F. (1996a). Monte Carlo slip inversion in the frequency domain: Application to the 1992 Nicaragua slow earthquakes. *Geophys. Res. Lett.* **23**, 913–916.
Ihmlé, P.F. (1996b). Frequency-dependent relocation of the 1992 Nicaragua slow earthquake: An empirical Green's function approach. *Geophys. J. Int.* **127**, 75–85.
Imran, J., Parker, G., Locat, J., Lee, H. (2001). A 1-D numerical model of muddy subaqueous and subaerial debris flows. *J. Hydraulic Eng.* **127**, 959–958.
Ishii, H., Abe, K. (1980). Propagation of tsunami on a linear slope between two flat regions. Part I edge wave. *J. Phys. Earth* **28**, 531–541.
Ishii, M., Shearer, P.M., Houston, H., Vidale, J.E. (2005). Extent, duration and speed of the 2004 Sumatra–Andaman earthquake imaged by the Hi-Net array. *Nature* **435**, 933–936.
Iverson, R.M., Denlinger, R.P. (2001). Flow of variably fluidized granular masses across three-dimensional terrain 1. Coulomb mixture theory. *J. Geophys. Res.* **106**, 537–552.
Janssen, T.T., Herbers, T.H.C., Battjes, J.A. (2008). Evolution of ocean wave statistics in shallow water: Refraction and diffraction over seafloor topography. *J. Geophys. Res.* **113**, doi:10.1029/2007JC004410.
Jeffreys, H., Jeffreys, B.S. (1956). Methods of Mathematical Physics. Third Edition, Cambridge University Press, Cambridge, England, 714 pp.
Jha, A.K., Winterstein, S.R. (2000). Nonlinear random ocean waves: Prediction and comparison with data, paper presented at Proceedings of the ETCE/OMAE2000 Joint Conference: Energy for the New Millennium, ASME, New Orleans, Louisiana, pp. 1–12.
Ji, C., Wald, D.J., Helmberger, D.V. (2002). Source description of the 1999 Hector Mine, California, earthquake, Part I: Wavelet domain inversion theory and resolution analysis. *Bull. Seismol. Soc. Am.* **92**, 1192–1207.
Jiang, L., Leblond, P.H. (1993). Numerical modeling of an underwater Bingham plastic mudslide and the waves which it generates. *J. Geophys. Res.* **98**, 10,303–310,317.
Johnson, J.M., Satake, K., Holdahl, S.R., Sauber, J. (1996). The 1964 Prince William Sound earthquake: Joint inversion of tsunami and geodetic data. *J. Geophys. Res.* **101**, 523–532.
Kajiura, K. (1963). The leading wave of a tsunami. *Bull. Earthquake Res. Inst.* **41**, 535–571.

Kajiura, K. (1970). Tsunami source, energy and the directivity of wave radiation. *Bull. Earthquake Res. Inst.* **48**, 835–869.

Kajiura, K. (1972). The directivity of energy radiation of the tsunami generated in the vicinity of a continental shelf. *J. Oceanogr. Soc. Japan* **28**, 260–277.

Kame, N., Rice, J.R., Dmowska, R. (2003). Effects of prestress state and rupture velocity on dynamic fault branching. *J. Geophys. Res.* **108(B5)**, ESE 13-11 - 13-21.

Kanamori, H. (1972). Mechanism of tsunami earthquakes. *Phys. Earth Planet. Int.* **6**, 346–359.

Kanamori, H., Given, J.W., Lay, T. (1984). Analysis of seismic body waves excited by the Mount St. Helens eruption of May 18, 1980. *J. Geophys. Res.* **89**, 1856–1866.

Kanamori, H., Kikuchi, M. (1993). The 1992 Nicaragua earthquake: A slow earthquake associated with subducted sediments. *Nature* **361**, 714–716.

Kanamori, H., Brodsky, E.E. (2004). The physics of earthquakes. *Rep. Prog. Phys.* **67**, 1429–1496.

Kanoglu, U., Synolakis, C.E. (1998). Long wave runup on piecewise linear topographies. *J. Fluid Mech.* **374**, 1–28.

Kennedy, A.B., Chen, Q., Kirby, J.T., Dalrymple, R.A. (2000). Boussinesq modeling of wave transformation, breaking and runup: I. one dimension. *Journal of Waterway, Port, Coastal, and Ocean Engineering* **126**, 39–47.

Kikuchi, M., Kanamori, H. (1991). Inversion of complex body waves–III. *Bull. Seismol. Soc. Am.* **81**, 2335–2350.

Kisslinger, C. (1993). Seismicity in subduction zones from local and regional network observations. *Pure Appl. Geophys.* **140**, 257–285.

Koshimura, S., Hayashi, Y., Munemoto, K., Imamura, F. (2008). Effect of the Emperor seamounts on trans-oceanic propagation of the 2006 Kuril Island earthquake tsunami. *Geophys. Res. Lett.* **35**, doi:10.1029/2007GL032129.

Kowalik, Z., Horrillo, J., Knight, W., Logan, T. (2008). Kuril Islands tsunami of November 2006: 1. Impact at Crescent City by distant scattering. *J. Geophys. Res.* **113**, doi:10.1029/2007JC004402.

Kreemer, C., Holt, W.E., Haines, A.J. (2002). The global moment rate distribution within plate boundary zones. In: Stein, S., Freymueller, J.T. (Eds.), Plate Boundary Zones. American Geophysical Union, Geodynamic Series, Washington, D.C, pp. 173–202.

La Rocca, M., Galluzzo, D., Saccorotti, G., Tinti, S., Cimini, G.B., Del Pezzo, E. (2004). Seismic signals associated with landslides and with a tsunami at Stromboli volcano, Italy. *Bull. Seismol. Soc. Am.* **94**, 1850–1867.

Lander, J.F. (1996). Tsunamis affecting Alaska 1737-1996, NGDC Key to Geophysical Records Documentation, 195 pp, U.S. Department of Commerce, National Oceanic and Atmospheric Administration, Boulder, Colorado.

Lavallée, D., Archuleta, R.J. (2003). Stochastic modeling of slip spatial complexities for the 1979 Imperial Valley, California, earthquake. *Geophys. Res. Lett.* **30**, doi:10.1029/2002GL015839.

Lavallée, D., Liu, P., Archuleta, R.J. (2006). Stochastic model of heterogeneity in earthquake slip spatial distributions. *Geophys. J. Int.* **165**, 622–640.

Lay, T., Wallace, T.C. (1995). Modern Global Seismology. Academic Press, San Diego, 517 pp.

Leblond, P.H., Mysak, L.A. (1978). Waves in the Ocean. Elsevier, Amsterdam, 602 pp.

Linker, M.F., Dieterich, J.H. (1992). Effects of variable normal stress on rock friction: Observations and constitutive equations. *J. Geophys. Res.* **97**, 4923–4940.

Liu, P.L.-F., Synolakis, C.E., Yeh, H. (1991). Report on the international workshop on long-wave run-up. *J. Fluid Mech.* **229**, 675–688.

Liu, P.L.-F. (2008). Tsunami modeling — propagation. In: Robinson, A.R., Bernard, E.N. (Eds.), The Sea. Harvard University Press, Cambridge, Massachusetts.

Liu, P.L.F., Yeh, H., Lin, P., Chang, K.T., Cho, Y.S. (1998). Generation and evolution of edge-wave packets. *Phys. Fluids* **10**, 1635–1657.

Liu-Zeng, J., Heaton, T.H., DiCaprio, C. (2005). The effect of slip variability on earthquake slip-length scaling. *Geophys. J. Int.* **162**, 841–849.

Locat, J. (1997). Normalized rheological behavior of fine mud and their flow properties in a pseudoplastic regime, in Debirs-Flow Hazards, mitigation, Mechanics, prediction, and Assessment, edited, pp. 260-269, Water Resources Engineering Division, American Society of Civil Engineers, New York.

Locat, J., Lee, H.J. (2002). Submarine landslides: Advances and challenges. *Canad. Geotech. J.* **39**, 193–212.
Longuet-Higgins, M.S. (1952). On the statistical distribution of the heights of sea waves. *J. Marine Res.* **11**, 245–266.
Lynett, P., Liu, P.L.F. (2002). A numerical study of submarine-landslide-generated waves and run-up. *Proc. R. Soc. Lond. A* **458**, 2885–2910.
Lynett, P. (2008). Modeling of tsunami inundation. In: Lee, W.H.K. (Ed.), Encyclopedia of Complexity and System Science. Springer-Verlag.
Lynett, P.J., Wu, T.-R., Liu, P.L.-F. (2002). Modeling wave runup with depth-integrated equations. *Coastal Eng.* **46**, 89–107.
Lynett, P.J., Liu, P.L.-F. (2005). A numerical study of run-up generated by three-dimensional landslides. *J. Geophys. Res.* **10**, doi:10.1029/2004JC002443.
Ma, K.-F., Satake, K., Kanamori, H. (1991). The origin of the tsunami excited by the 1989 Loma Prieta earthquake-faulting or slumping. *Geophys. Res. Lett.* **18**, 637–640.
Ma, X.Q., Kusznir, N.J. (1994). Effects of rigidity layering, gravity and stress relaxation on 2-D subsurface fault displacement fields. *Geophys. J. Int.* **118**, 201–220.
Madsen, P.A., Schäffer, H.A. (1998). Higher-order Boussinesq-type equations for surface gravity waves: Derivation and analysis. *Phil. Trans. R. Soc. London, A* **356**, 3123–3184.
Madsen, P.A., Fuhrman, D.R., Schäffer, H.A. (2008). On the solitary wave paradigm for tsunamis. *J. Geophys. Res.* **113**, doi:10.1029/2008JC004932.
Mai, P.M., Beroza, G.C. (2002). A spatial random field model to characterize complexity in earthquake slip. *J. Geophys. Res.* **107**, doi:10.1029/2001JB000588.
Martel, S.J. (2004). Mechanics of landslide initiation as a shear fracture phenomenon. *Mar. Geol.* **203**, 319–339.
Massel, S.R. (1996). Ocean Surface Waves: Their Physics and Prediction. World Scientific, Singapore, 508 pp.
Masterlark, T., DeMets, C., Wang, H.F., Sánchez, O., Stock, J. (2001). Homogeneous vs heterogeneous subduction zone models: Coseismic and postseismic deformation. *Geophys. Res. Lett.* **28**, 4047–4050.
Masterlark, T. (2003). Finite element model predictions of static deformation from dislocation sources in a subduction zone: Sensitivities to homogeneous, isotropic, Poisson-solid, and half-space assumptions. *J. Geophys. Res.* **103**, doi:10.1029/2002JB002296.
McCaffrey, R. (1994). Dependence of earthquake size distributions on convergence rates at subduction zones. *Geophys. Res. Lett.* **21**, 2327–2330.
Mei, C.C. (1989). The Applied Dynamics of Ocean Surface Waves. World Scientific, Singapore, 740 pp.
Meltzner, A., Sieh, K., Abrams, M., Agnew, D.C., Hudnut, K.W., Avouac, J.P., Natawidjaja, D. (2006). Uplift and subsidence associated with the Great Aceh-Andaman earthquake of 2004. *J. Geophys. Res.* **111**, doi:10.1029/2005JB003891.
Mendoza, C., Fukuyama, E. (1996). The July 12, 1993, Hokkaido-Nansei-Oki, Japan, earthquake: Coseismic slip pattern from strong-motion and teleseismic recordings. *J. Geophys. Res.* **101**, 791–801.
Michael, A.J., Geller, R.J. (1984). Linear moment tensor inversion for shallow thrust earthquakes combining first-motion and surface wave data. *J. Geophys. Res.* **89**, 1889–1897.
Mikada, H., Mitsuzawa, K., Matsumoto, H., Watanabe, K., Morita, S., Otsuka, R., Sugioka, H., Baba, T., Araki, E., Suyehiro, K. (2006). New discoveries in dynamics of an M8 earthquake-phenomena and their implications from the 2003 Tokachi–Oki earthquake using a long term monitoring cabled observatory. *Tectonophys.* **426**, 95–105.
Miller, G.R., Munk, W.H., Snodgrass, F.E. (1962). Long-period waves over California's continental borderland Part II. Tsunamis. *J. Marine Res.* **20**, 31–41.
Miyazaki, S., Larson, K.M., Choi, K., Hikima, K., Koketsu, K., Bodin, P., Haase, J., Emore, G., Yamagiwa, A. (2004). Modeling the rupture process of the 2003 September 25 Tokachi–Oki (Hokkaido) earthquake using 1-Hz GPS data. *Geophys. Res. Lett.* **31**, doi:10.1029/2004GL021457.
Mofjeld, H.O., González, F.I., Bernard, E.N., Newman, J.C. (2000a). Forecasting the heights of later waves in Pacific-wide tsunamis. *Nat. Haz.* **22**, 71–89.
Mofjeld, H.O., Titov, V.V., González, F.I., Newman, J.C. (2000b). Analytic theory of tsunami wave scattering in the open ocean with application to the North Pacific Ocean, NOAA Technical Memorandum OAR PMEL-116, 38 pp.

Mofjeld, H.O., Titov, V.V., González, F.I., Newman, J.C. (2001). Tsunami scattering provinces in the Pacific Ocean. *Geophys. Res. Lett.* **28**, 335–337.

Mofjeld, H.O., Symons, C.M., Lonsdale, P., González, F.I., Titov, V.V. (2004). Tsunami scattering and earthquake faults in the deep Pacific Ocean. *Oceanography* **17**, 38–46.

Mofjeld, H.O. (2009). Tsunami measurements. In: Bernard, E.N., Robinson, A.R. (Eds.), The Sea. Harvard University Press, Cambridge, MA, pp. 201–235.

Moore, G.F., Bangs, N.L., Taira, A., Kuramoto, S., Pangborn, E., Tobin, H.J. (2007). Three-dimensional splay fault geometry and implications for tsunami generation. *Science* **318**, 1128–1131.

Munk, W., Snodgrass, F.E., Gilbert, F. (1964). Long waves on the continental shelf: An experiment to separate trapped and leaky modes. *J. Fluid Mech.* **20**, 529–554.

Muraleedharan, G., Rao, A.D., Kurup, P.G., Nair, N.U., Sinha, M. (2007). Modified Weibull distribution for maximum and significant wave height simulation and prediction. *Coastal Eng.* **54**, 630–638.

Mysak, L.A. (1978). Wave propagation in random media, with oceanic applications. *Rev. Geophys. Space Phys.* **16**, 233–261.

Nettles, M. (2007). Analysis of the 10 February 2006: Gulf of Mexico earthquake from global and regional seismic data, paper presented at 2007 Offshore Technology Conference, Houston, Texas, p. 2.

Nielsen, S.B., Carlson, J.M. (2000). Rupture pulse characterization: Self-healing, self-similar, expanding solutions in a continuum model of fault dynamics. *Bull. Seismol. Soc. Am.* **90**, 1480–1497.

Nishigami, K. (1991). A new inversion method of coda waveforms to determine spatial distribution of coda scatterers in the crust and uppermost mantle. *Geophys. Res. Lett.* **18**, 2225–22228.

Nishigami, K., Matsumoto, S. (2008). Imaging inhomogeneous structures in the earth by coda envelope inversion and seismic array observation. *Adv. Geophys.* **50**, 301–318.

Nosov, M.A., Kolesov, S.V. (2007). Elastic oscillations of water column in the 2003 Tokachi–Oki tsunami source: In-situ measurements and 3-D numerical modeling. *Natural Hazards and Earth System Sciences* **7**, 243–249.

Novikova, T., Wen, K.-L., Huang, B.-S. (2002). Analytical model for gravity and Rayleigh -Wave investigation in the layered ocean-Earth structure. *Bull. Seismol. Soc. Am.* **92**, 723–738.

Oglesby, D.D., Day, S.M. (2002). Stochastic fault stress: Implications for fault dynamics and ground motion. *Bull. Seismol. Soc. Am.* **92**, 3006–3021.

Ohmachi, T., Tsukiyama, H., Matsumoto, H. (2001). Simulation of tsunami induced by dynamic displacement of seabed due to seismic faulting. *Bull. Seismol. Soc. Am.* **91**, 1898–1909.

Okal, E.A. (1988). Seismic parameters controlling far-field tsunami amplitudes: A review. *Nat. Haz.* **1**, 67–96.

Okal, E.A. (2003). Normal mode energetics for far-field tsunamis generated by dislocations and landslides. *Pure Appl. Geophys.* **160**, 2189–2221.

Okal, E.A., Synolakis, C.E. (2004). Source discriminants for near-field tsunamis. *Geophys. J. Int.* **158**, 899–912.

Park, J.-O., Tsuru, T., Kodiara, S., Cummins, P.R., Kaneda, Y. (2002). Splay fault branching along the Nankai subduction zone. *Science* **297**, 1157–1160.

Pelinovsky, E.N., Mazova, R.K. (1992). Exact analytical solutions of nonlinear problems of tsunami wave run-up on slopes with different profiles. *Nat. Haz.* **6**, 227–249.

Petley, D.N., Higuchi, T., Petley, D.J., Bulmer, M.H., Carey, J. (2005). Development of progressive landslide failure in cohesive materials. *Geology* **33**, 201–204.

Piatanesi, A., Lorito, S. (2007). Rupture process of the 2004 Sumatra–Andaman earthquake from tsunami waveform inversion. *Bull. Seismol. Soc. Am.* **97**, S223–S231.

Piper, D.J.W., Cochonat, P., Morrison, M.L. (1999). The sequence of events around the epicentre of the 1929 Grand Banks earthquake: Initiation of debris flows and turbidity current inferred from sidescan sonar. *Sedimentology* **46**, 79–97.

Polet, J., Kanamori, H. (2000). Shallow subduction zone earthquakes and their tsunamigenic potential. *Geophys. J. Int.* **142**, 684–702.

Polet, J., Thio, H.K. (2003). The 1994 Java tsunami earthquake and its "normal" aftershocks. *Geophys. Res. Lett.* **30**, doi:10.1029/2002GL016806.

Poliakov, A.N.B., Dmowska, R., Rice, J.R. (2002). Dynamic shear rupture interactions with fault bends and off-axis secondary faulting. *J. Geophys. Res.* **107**, doi:10.1029/2001JB000572.

Pollitz, F.F., Banerjee, P., Grijalva, K., Nagarajan, B., Bürgmann, R. (2008). Effect of 3-D viscoelastic structure on post-seismic relaxation from the 2004 M = 9.2 Sumatra earthquake. *Geophys. J. Int.* **173**, 189–204.
Power, W.L., Tullis, T.E. (1991). Euclidean and fractal models for the description of rock surface roughness. *J. Geophys. Res.* **96**, 415–424.
Prevosto, M., Kogstad, H.E., Robin, A. (2000). Probability distributions for maximum wave and crest heights. *Coastal Eng.* **40**, 329–360.
Rabinovich, A.B. (1997). Spectral analysis of tsunami waves: Separation of source and topography effects. *J. Geophys. Res.* **102**, 12,663–612,676.
Rabinovich, A.B., Stephenson, F.E., Thomson, R.E. (2006). The California tsunami of 15 June 2005 along the coast of North America. *Atmosphere-Ocean* **44**, 415–427.
Rabinovich, A.B., Thomson, R.E. (2007). The 26 December 2004 Sumatra tsunami: Analysis of tide gauge data from the world ocean Part 1. Indian Ocean and South Africa. *Pure Appl. Geophys.* **164**, 261–308.
Rayleigh, J.W.S. (1880). On the resultant of a large number of vibrations of the same pitch and of arbitrary phase. *Phil. Mag.* **10**, 73–78.
Rhie, J., Dreger, D.S., Bürgmann, R., Romanowicz, B. (2007). Slip of the 2004 Sumatra–Andaman earthquake from joint inversion of long-period global seismic waveforms and GPS static offsets. *Bull. Seismol. Soc. Am.* **97**, S115–S127.
Rice, J.R. (1983). Constitutive relations for fault slip and earthquake instabilities. *Pure Appl. Geophys.* **121**, 443–475.
Rice, J.R., Cocco, M. (2007). Seismic fault rheology and earthquake dynamics. In: Handy, M.R., et al. (Eds.), Tectonic Faults: Agents of Change on a Dynamic Earth. MIT Press, Cambridge, MA, pp. 99–137.
Rudnicki, J.W., Wu, M. (1995). Mechanics of dip-slip faulting in an elastic half-space. *J. Geophys. Res.* **22**, 173–22186.
Ruff, L., Kanamori, H. (1983). Seismic coupling and uncoupling at subduction zones. *Tectonophys.* **99**, 99–117.
Ruina, A.L. (1983). Slip instability and state variable friction laws. *J. Geophys. Res.* **88**, 10,359–310,370.
Rybicki, R. (1986). Dislocations and their geophysical application. In: Teisseyre, R. (Ed.), Continuum Theories in Solid Earth Physics. Elsevier, Amsterdam, pp. 18–186.
Rychlik, I., Johannesson, P., Leadbetter, M.R. (1997). Modelling and statistical analysis of ocean-wave data using transformed Gaussian processes. *Marine Struct.* **10**, 13–47.
Rychlik, I., Leadbetter, M.R. (1997). Analysis of ocean waves by crossing and oscillation intensities, paper presented at Proceedings of the Seventh International Offshore and Polar Engineering Conference (ISOPE), Golden, Colorado, pp. 206–213.
Sagiya, T., Thatcher, W. (1999). Coseismic slip resolution along a plate boundary megathrust: the Nankai Trough, southwest Japan. *J. Geophys. Res.* **104**, 1111–1129.
Saito, T., Furumura, T. (2009a). Scattering of linear long-wave tsunamis due to randomly fluctuating sea-bottom topography: Coda excitation and scattering attenuation. *Geophys. J. Int.* **177**, 958–965.
Saito, T., Furumura, T. (2009b). Three-dimensional tsunami generation simulation due to sea-bottom deformation and its interpretation based on linear theory. *Geophys. J. Int.* **178**, 877–888.
Saito, T., Furumura, T. (2009c). Three-dimensional simulation of tsunami generation and propagation: Application to intraplate events. *J. Geophys. Res.* **114**, doi:10.1029/2007JB005523.
Samorodnitsky, G., Taqqu, M.S. (1994). Stable Non-Gaussian Random Processes: Stochastic Models with Infinite Variance. Chapman & Hall/CRC Press, Boca Raton, Florida, 632 pp.
Satake, K., Tanioka, Y. (1999). Sources of tsunami and tsunamigenic earthquakes in subduction zones. *Pure Appl. Geophys.* **154**, 467–483.
Sato, K., Minagawa, N., Hyodo, M., Baba, T., Hori, T., Kaneda, Y. (2007). Effect of elastic inhomogeneity on the surface displacements in the northeastern Japan: Based on three-dimensional numerical modeling. *Earth Planets Space* **59**, 1083–1093.
Savage, J.C. (1998). Displacement field for an edge dislocation in a layered half-space. *J. Geophys. Res.* **103**, 2439–2446.
Scholz, C.H. (1998). Earthquakes and friction laws. *Nature* **391**, 37–42.

Seno, T. (2002). Tsunami earthquakes as transient phenomena. *Geophys. Res. Lett.* **29**, doi:10.1029/2002GL014868.

Seno, T., Hirata, K. (2007). Is the 2004 Sumatra–Andaman earthquake a tsunami earthquake? *Bull. Seismol. Soc. Am.* **97**, S296–S306.

Shapiro, S.S., Wilk, M.B. (1965). An analysis of variance test for normality (complete samples). *Biometrika* **52**, 591–611.

Shuto, N., Chida, K., Imamura, F. (1995). Generation mechanism of the first wave of the 1983 Nihonkai–Chubu earthquake tsunamis. In: Tsuchiya, Y., Shuto, N. (Eds.), Tsunami: Progress in Prediction, Disaster Prevention and Warning. Kluwer Academic Publishers, Dordrecht, pp. 37–53.

Shuto, N., Matsutomi, H. (1995). Field survey of the 1993 Hokkaido Nansei-Oki earthquake tsunami. *Pure Appl. Geophys.* **144**, 649–663.

Sibuet, J.-C., Rangin, C., Le Pichon, X., Singh, S., Cattaneo, A., Graindorge, D., Klingelhoefer, F., Lin, J.-Y., Malod, J., Maury, T., Schneider, J.-L., Sultan, N., Umber, M., Yamaguchi, H., Sumatra Aftershocks Team (2007). 26th December 2004 great Sumatra–Andaman earthquake: Co-seismic and post-seismic motions in northern Sumatra. *Earth Planet. Sci. Lett.* **263**, 88–103.

Singh, S.J., Punia, M., Rani, S. (1994). Crustal deformation due to non-uniform slip along a long fault. *Geophys. J. Int.* **118**, 411–427.

Snodgrass, F.E., Munk, W.H., Miller, G.R. (1962). Long-period waves over California's continental borderland Part I. Background spectra. *J. Marine Res.* **20**, 3–30.

Somerville, P., Irikura, K., Graves, R., Sawada, S., Wald, D., Abrahamson, N.A., Iwasaki, Y., Kagawa, T., Smith, N., Kowada, A. (1999). Characterizing crustal earthquake slip models for the prediction of strong ground motion. *Seismol. Res. Lett.* **70**, 59–80.

Stephens, M.A. (1974). EDF statistics for goodness of fit and some comparisons. *J. Amer. Statist. Assoc.* **69**, 730–737.

Stoker, J.J. (1957). Water Waves: The Mathematical Theory with Applications. Interscience Publishers, Inc., New York, 567 pp.

Subarya, C., Chlieh, M., Prawirodirdjo, L., Avouac, J.P., Bock, Y., Sieh, K., Meltzner, A., Natawidjaja, D., McCaffrey, R. (2006). Plate-boundary deformation associated with the great Sumatra–Andaman earthquake. *Nature* **440**, 46–51.

Synolakis, C.E. (1987). The runup of solitary waves. *J. Fluid Mech.* **185**, 523–545.

Tadepalli, S., Synolakis, C.E. (1994). The run-up of N-waves on sloping beaches. *Proc. R. Soc. Lond. A* **445**, 99–112.

Takahara, M., Yomogida, K. (1992). Estimation of Coda Q using the maximum likelihood method. *Pure Appl. Geophys.* **139**, 255–268.

Tanioka, Y., Ruff, L., Satake, K. (1995). The great Kuril earthquake of October 4, 1994 tore the slab. *Geophys. Res. Lett.* **22**, 1661–1664.

Tanioka, Y., Ruff, L.J., Satake, K. (1997). What controls the lateral variation of large earthquake occurrence along the Japan trench? *The Island Arc* **6**, 261–266.

Tanioka, Y., Nishimura, Y., Hirakawa, K., Imamura, F., Abe, I., Abe, Y., Shindou, K., Matsutomi, H., Takahashi, T., Imai, K., Harada, K., Namegawa, Y., Hasegawa, Y., Hayashi, Y., Nanayama, F., Kamataki, T., Kawata, Y., Fukasawa, Y., Koshimura, S., Hada, Y., Azumai, Y., Hirata, K., Kamikawa, A., Yoshikawa, A., Shiga, T., Kobayashi, M., Masakam, S. (2004). Tsunami run-up heights of the 2003 Tokachi-Oki earthquake. *Earth Planets Space* **56**, 359–365.

Taylor, M.A.J. (1998). Chapter VI, Friction, dilatancy, and pore fluids in controlling subduction earthquake rupture. PhD Thesis. Harvard University Cambridge, MA, 41 pp.

Taylor, M.A.J., Dmowska, R., Rice, J.R. (1998). Upper plate stressing and seismicity in the subduction earthquake cycle. *J. Geophys. Res.* **103**, 24,523–524,542.

Templeton, E.L., Baudet, A., Bhat, H.S., Dmowska, R., Rice, J.R., Rosakis, A.J., Rousseau, C.-E. (2009). Finite element simulations of dynamic shear rupture experiments and dynamic path selection along kinked and branched faults. *J. Geophys. Res.* **114**, doi:10.1029/2008JB006174.

The WAFO Group, (2000). WAFO: A Matlab Toolbox for Analysis of Random Waves and Loads, Tutorial Version 2.0.02. Lund Institute of Technology, Lund University, Lund, Sweden. 124 pp.

Thio, H.K., Graves, R.W., Somerville, P.G., Sato, T., Ishii, T. (2004). A multiple time window rupture model for the 1999 Chi–Chi earthquake from a combined inversion of teleseismic, surface wave, strong motion, and GPS data. *J. Geophys. Res.* **109**, doi:10.1029/2002JB002381.

Tinti, S., Tonini, R. (2005). Analytical evolution of tsunamis induced by near-shore earthquakes on a constant-slope ocean. *J. Fluid Mech.* **535**, 33–64.
Trifunac, M.D., Hayir, A., Todorovska, M.I. (2002). Was Grand Banks event of 1929 a slump spreading in two directions? *Soil Dynam. Earthquake Eng.* **22**, 349–360.
Tsai, C.P. (1997a). Slip, stress drop and ground motion of earthquakes: A view from the perspective of fractional Brownian motion. *Pure Appl. Geophys.* **149**, 689–706.
Tsai, C.P. (1997b). Ground motion modeling for seismic hazard analysis in the near-source regime: An asperity model. *Pure Appl. Geophys.* **149**, 265–297.
Tsai, V.C., Nettles, M., Ektröm, G., Dziewonski, A.M. (2005). Multiple CMT source analysis of the 2004 Sumatra earthquake. *Geophys. Res. Lett.* **32**, doi:10.1029/2005GL023813.
Ursell, F. (1952). Edge waves on a sloping beach. *Proc. R. Soc. Lond. A* **214**, 79–97.
van Dorn, W.G. (1984). Some tsunami characteristic deducible from tide records. *J. Phys. Oceanogr.* **14**, 353–363.
van Dorn, W.G. (1987). Tide gage response to tsunamis. Part II: Other oceans and smaller seas. *J. Phys. Oceanogr.* **17**, 1507–1517.
Varnes, D.J. (1978). Slope movement types and processes. In: Schuster, R.L., Krizek, R.J. (Eds.), Landslides: Analysis and Control. National Academy of Sciences, Washington, D.C, pp. 11–33.
Ward, S.N. (1982). On tsunami nucleation II. An instantaneous modulated line source. *Phys. Earth Planet. Int.* **27**, 273–285.
Watanabe, H. (1972). Statistical studies on the wave-form and maximum height of large tsunamis. *J. Oceanogr. Soc. Japan* **28**, 229–241.
Wells, D.L., Coppersmith, K.J. (1994). New empirical relationships among magnitude, rupture length, rupture width, rupture area, and surface displacement. *Bull. Seismol. Soc. Am.* **84**, 974–1002.
Wendt, J., Oglesby, D.D., Geist, E.L. (2009). Tsunamis and splay fault dynamics. *Geophys. Res. Lett.*, doi:10.1029/2009GL038295.
Wessel, P., Smith, W.H.F. (1995). New version of the Generic Mapping Tools released. *Eos, Trans. Am. Geophys. U.* **76**, F329.
Wu, T.Y. (1981). Long waves in ocean and coastal waters. *J. Eng. Mech. Division, ASCE* **107**, 501–522.
Yagi, Y. (2004). Source rupture process of the 2003 Tokachi–Oki earthquake determined by joint inversion of teleseismic body wave and strong ground motion data. *Earth Planets Space* **56**, 311–316.
Yamanaka, Y., Kikuchi, M. (2003). Source process of the recurrent Tokachi–Oki earthquake on September 26, 2003 inferred from teleseismic body waves. *Earth Planets Space* **55**, e21–e24.
Yeh, H. (2008). Tsunami impacts on coastlines. In: Robinson, A.R., Bernard, E.N. (Eds.), The Sea. Harvard University Press, Cambridge, Massachusetts, pp. 333–369.
Yeh, H.H. (1991). Tsunami bore runup. *Nat. Haz.* **4**, 209–220.
Yoshioka, S., Hashimoto, M., Hirahara, K. (1989). Displacement fields due to the 1946 Nankaido earthquake in a laterally inhomogeneous structure with the subducting Philippine Sea plate–a three-dimensional finite element approach. *Tectonophys.* **159**, 121–136.
Zhao, S., Müller, R.D., Takahashi, Y., Kaneda, Y. (2004). 3-D finite-element modeling of deformation and stress associated with faulting: Effect of inhomogeneous crustal structures. *Geophys. J. Int.* **157**, 629–644.
Zheng, G., Rice, J.R. (1998). Conditions under which velocity-weakening friction allows a self-healing versus a cracklike mode of rupture. *Bull. Seismol. Soc. Am.* **88**, 1466–1483.

Index

A

Acceleration, 13
　average slip, 36
　peak, 14
Accelerometers
　borehole triaxial, 3
　triaxial, 5, 25
　uniaxial, 4
Accretionary wedge, 127
Acoustic emission (AE)
　events of, 12–14, 38
　experiments on, 37
　systems of, 11, 44
Active fields, 68
After-slip, 121
Aftershock, 90, 91. *See also* Main shock
Aftershocks, 10, 11
　sequences of, 12, 13
　spatial distribution of, 34, 42
Along-strike rupture
　complexity of, 132
Ambient noise, 71
Ambient stress, 3, 15, 24
Analysis, 148
　broadside, 130
　1D, 110
　dimensional, 75, 91
　fatigue, 143
　probabilistic seismic hazard, 95
　seismic coda, 140
　statistical, 93
　trapped and shelf modes, 152
Analytical techniques, 78
Anelastic effects, 99
Angle
　branch, 127
　incidence, 149
　slope, 116
　steep take-off, 120
Anisotropy, 23
Anticline, 91
　dome structure, 84
　elongated, 77
Apex
　upward-directed, 87
Apollo-Hendrick, 60
　clustering, 68

Apparent stress
　constant attenuation models, 35
　plot of, 36
　range of, 24
　ratio, 34, 39
　roof fall events, 20
　static stress drop, 37
　stress release, 32
　systematic dependence, 36
　upper bound of, 40
　various mines, 44
Arrival
　first, 6
　P- and *S*-, 5
　P-wave first, 40
Aseismic, 77, 85, 96
Asperity, 85
Assessment
　rock mass damage, 14
　seismic hazard, 3
　seismic issues, 2
Assumptions
　fault geometry, 120
　narrow-band, 152
　uniform thickness, 127
Attenuation, 118, 125, 156
Attenuation coefficient, 149
Attenuation factor
　seismic, 141
Australia
　Darlot Gold Mine, 19
　DD location method, 6
　monitoring systems, 4
　seismic monitoring in, 1, 3, 43
Azimuthal anisotropy, 66, 67
Azimuths
　strike-normal, 128, 145

B

Basin-wide propagation, 107
Bathymetric data
　high-resolution, 108
Bathymetry, 152, 157
　changes in, 140
　continental shelf, 145
　nearshore, 108
　ocean, 149
　offshore, 130

Beaming effect, 145
Beaming pattern
　primary, 129
Bergermeer, 58
　active field, 68
　geometry, 87
Best-located events
　epicenter map of, 9
Better-located events
　epicenter map of, 8, 9
Bilinear rheology, 116
Biot
　pore pressure coefficient, 80
　theory, 78
24-bit resolution ESG, 5
Blind deconvolution, 32
Body waves
　amplitude spectra of, 22
Bottom Pressure Recorder (BPR), 125
　advent of, 140
　DART, 158
　North Pacific map, 140
　tsunami, 137
Brine, 71
Bukhara-Gissar, 92
Bulk modulus, 80
Burger's vector, 123
　solutions of, 124
Bursting
　incipient pillar, 25

C

Cajon Pass, 34
Carletonville, 2
Cascadia
　slip on, 139
　strike of, 152
Catalog completeness
　threshold of, 8
Chalk
　formation of, 71
　pressured, 78
Chalk reservoir, 77
Chaotic phenomenon, 68
China
　Laohutai coal mine, 43
　mining operations in, 4
　seismic monitoring in, 1, 3

Classic joint hypocenter, 6
Clustering, 9
Clusters
 AE events, 11, 13
 Oryx Gold Mine, 22
 plane view of, 12
Coal bumps, 4
Coast Range-Sierran Block, 91
Coastal interaction, 150–156
Cogdell Canyon Reef, 61
 magnitude 4.6-4.7, 75
 temporal patterns, 64
Collapsing, 72
 illustration of, 66
Compaction, 71
 differential, 85
 map of, 99
 reservoir, 99
Complex reservoir, 96
Components, 22–24
 compliant, 41
 isotropic, 26, 27
 positive volumetric, 25
Concentration at the field center
 production wells, 83
 surface subsidence, 87
Constitutive Parameters
 additional, 112
 determination of, 111
 estimation of, 158
 slip dynamics, 113
Contact
 gas-water, 93
 inter-grain, 78
Continuity equation, 116, 117
Contracted center, 87
Controlled experiments, 73, 74
Convergence
 trench-normal, 112
Convolution, 81, 82
Coordinate transformation
 hodograph, 118
Corner frequency, 30
 cubed, 35
 estimation of, 36
Correlation with production, 57, 62
Coseismic slip
 inversion, 125, 156
 model, 149, 157, 158
 wavenumber, 121–123, 134
Costa Oriental, 61, 82, 96
Coulomb failure criterion, 83
Crack growth, 77

Cracks, 67
Creighton Mine
 crown pillar at, 43
 data from, 14, 18
 space-time correlation, 42
Crest
 amplitude of, 143, 153
 wave, 156
Critical earthquakes
 concept of, 1, 3, 45
Critical points, 42
Critical rupture, 43
Crustal earthquakes, 25
Crustal stress
 ambient state of, 15
Crustal unloading, 93
CSIR, 4
CSIRO, 4
CTBT, 3
Cumulative frequency
 b values, 17, 18
 ratio values, 16

D

Dams, 55
Darlot Gold Mine, 19
Data. See also 4D
 high-quality, 96, 98
 seismic, 62, 99
 time lapse, 96
DD location method, 6. See also
 Double-difference
 relocated by, 7, 8
Debris avalanches
 subaerial, 116
 submarine, 115, 117
Deep mines
 events originating in, 3
 Kolar Gold Fields, 5
 loss of life at, 4
 research programme, 17
 seismic events in, 20
 South African, 2, 38, 40, 42, 44, 45
Deep tabular mines, 21
DEEPMINE Collaborative
 Research, 17
Deformation, 77, 91. See also
 permanent
 brittle, 87
 central bulge of, 83
 elastic, 113
 horizontal, 125
 mechanical, 78

nonlinear, 98
 seafloor, 114
 surface, 80, 88
 vertical seafloor, 111
 wells and casing, 71
Depletion strategy, 93
Depths
 focal, 90
 hypocentral, 89
 midcrustal, 88, 91
 reservoir, 86
Design
 deep-level stopes, 15
 mine, 2, 18, 43, 44
Destructive earthquake sequences, 55, 92
 Coalinga, 89–91
 Gazli, 91–93, 95
Deterministic, 107
 analyses, 110
 approach, 157
 scattering index, 149
Deviatoric mechanisms, 12, 27
 pure, 24
DIANA
 software package, 86
Dilatation
 center of, 81
 ring of, 82
Directivity effects, 27, 30, 44
Dislocation model
 uniform-slip, 110
Dispersive coda, 144
Displacement, 80, 81. See also
 monopole
 dipole, 147
 horizontal, 88
 radial, 83, 84
 shear, 86
 static, 113, 123, 139, 144
 vertical, 82
Distribution model
 Gaussian, 142, 154
 Rayleigh cumulative, 144, 154
 slip, 121, 122
Dome structure, 83, 84
Dong Tan, 4
Double-couple, 65
Double-couple events, 26
Double-difference (DD), 6, 7
Doublets
 seismic, 11
Dutch law, 68, 95
Dykes

Index

deep mines, 20
experimental field without, 2
identification of, 17
strong, 3
Dynamic processes
solid-earth, 109

E

Earthquake
dynamics, 113, 157
focal mechanism, 120
friction, 112, 117, 127, 158
inversion (slip), 125, 156
kinematic, 110, 114, 120, 147
rupture, 108, 110, 112, 124, 128, 129, 132, 134, 145, 146
seismic moment, 111, 135, 136, 157
spectrum, 118, 122, 123, 143, 152, 154, 155
static, 111, 121, 123, 139, 144
waveforms, 108, 118, 121, 130, 138, 140, 148, 150
Earthquakes,. See tsunamis
Efficiency. See also Savage-Wood
conventional seismic, 39
radiation, 31, 44
Eigenmodes
coastal, 152
Ekofisk, 59
compaction map, 99
microseismic monitoring project, 71
microseismicity, 72
moderate event, 96
reservoir management, 77
seismicity map of, 62, 63
sophisticated modeling, 98
subsidence, 57, 82
El Teniente Mine, 19
Elastic
deformation, 113
dislocations, 130
displacement, 123, 144, 158
homogeneous, 126
moduli, 111
response, 135
static, 114, 125, 139
Elastic wave theory, 113
Elliptical disturbance, 137
normal and parallel, 137
Empirical Green's Function, 28
convolution, 33

deconvolution technique, 44
source parameters, 35
Energy, 145, 152
accumulated, 78
class of, 92
coda, 149, 156
edge wave, 138
high-frequency, 127
potential, 109
tsunami, 142
Energy flux
conservation of, 134
Engineering Seismology Group (ESG)
24-bit resolution, 5
MS and AE, 11
Enhanced recovery, 68
Environment
fluid injection, 77
hard-rock, 71
sedimentary, 71
thrust faulting, 83
Epicenters, 62–64, 66
Cogdell, 75
events, 83, 94
Groningen field, 68
map of, 76
microseismicity, 72
Escarpments, 148
Eshelby inclusion, 99
European Seismological Commission, 65
Event, 8
acoustic emission, 38
automatic, 6
Bernburgat, 22
better-located, 9
comparison for, 31
crack opening and closure, 12
damaging, 19
Driefontein Gold Mine, 16
higher energy, 44
implosional, 24
Klerksdorp, 4
large, 2, 29, 45
locations of, 25
low frequency, 27
main, 10
master, 7
Matjhabeng, 21
microseismic, 15
mining tectonic, 9
mining-induced, 3
moment tensors of, 26

MS and AE, 12–14
multiple, 11
observed, 8
prediction of, 40–43
probability density function (PDF), 7
pure shear, 25
records of selected, 30
roof fall, 20
seismic, 1, 13, 14, 23, 35
single, 7
small, 17, 28, 35
Solikamsk-2 mine, 5
static stress drop for, 37
STF for, 31, 33
sub, 27
Event localization, 71, 72
seismicity map, 94
Excitation
edge waves, 107, 118, 137, 138, 152, 154, 157
trapped modes, 130, 150
Experiments
acoustic emission, 37
earthquake-generation, 40
locations, 6
seismic monitoring, 4
semi-controlled seismogenic, 2
Exponential curve
best-fit, 141
Extended ring, 87

F

Factors inducing seismic activity, 57, 63–65
Failure mechanism, 20
Far field, 138–156
Fashing, 61
hypothesis, 77
Pennington, 86
pre-existing fault, 63
Fault geometry, 113
assumptions of, 120
non-planar, 139
Fault plane
geodetic, 92
Faulting, 64, 88
analog models of, 87
thrust, 83
Faults. See also outer rise
activation of, 64, 98
active, 20
average size of, 42

cone-shaped, 87
Dagbreek, 21
density, 69
experimental field without, 2
formation of new, 65
friction of, 15
identification of, 17
inter-plate thrust, 109, 112, 120, 123, 134, 139, 147
length of, 30, 37
local, 27
locking of, 57
major, 25, 26
mapping of, 71
movement of, 4
normal, 69, 86
pattern of, 71
planes of, 3
pre-existing, 24
radius of, 34
reactivation of, 58
reverse, 86, 87
roughness, 121, 122, 149
scarps, 113, 115, 124, 125
seismic, 26
shearing on, 43
slip direction of, 74
slip of, 18
splay, 127, 158
steep reverse, 87
strains on, 3
strike slip, 111, 152
trapping, 86
trending, 68
1991 field survey, 94
Finite-time singularity, 43
Finiteness factor, 145
First motion
　compressional and dilatational, 74
Flow conduits, 73
Fluid
　distributions of, 41
　extraction, 56, 65, 73, 79–83, 85–91, 93
　injection, 57, 58, 73, 75–77, 98, 99
　injection of, 2
　mapping of, 71, 96
　migration of, 56
　overpressured, 68
　pore, 15
Fluid movement, 68, 71
Focal plane solutions, 74

Folds, 90, 92
Footwall
　quartzite, 16
　shale, 15
Foreshocks, 10, 11
　concentrated, 40, 44
　sequences of, 13
Formation
　new faults, 65
　soft chalk, 71
Formula
　time-to-failure, 42
Fourier transform, 122
Fracture-dominated rupture, 7
Fractures, 56, 71, 73
Friction law
　earthquake, 113
　rate- and state-, 112
Friction-dominated slip, 7

G

Gas cloud, 71, 72
Gas reservoir, 64, 65
　on-land, 68
Gaussian function, 148
Gaussian null hypothesis, 142, 152
Geochemical process, 77, 78, 86
Geodetic. *See also* geodetic
　inversions, 120, 121
　data, 120
　measurements, 121
Geomechanical models
　2D and 3D, 86
Geometrical effects
　impact of, 86
Geophones, 71, 72
　CSIRO, 4
　surface, 5
　three-component, 26
Geothermal fields, 55, 94
Gobles, 58, 63, 75
Goose Creek, 61
　legal consequences, 97
　new faults, 65
　subsidence, 78, 96
Gouge, 25
　fault, 21
Gouge material, 20
Gouge zone, 21
Granite
　seismic response in, 21
Granodiorite
　homogeneous, 21

Granular theory, 115. *See also* mixture
Green's function, 23, 32
　displacement, 81
　elastic, 114, 125
　elastostatic, 80
　empirical, 28, 33, 35, 44
　theoretical, 22
　tsunami, 134, 135, 152
　vertical, 82
Groningen, 58
　active field, 68
　geometry, 87
　induced seismicity, 69
　surface subsidence, 96
Ground sand
　undercompacted, 87
Ground-motion estimates, 14
Ground-motion parameters
　near-source, 13
　observations of, 14
Grozny, 60
　seismicity at, 78
Gulf of Alaska, 130, 155
Gullfaks reservoir, 96, 98
Gutenberg-Richter
　diagram of, 7
　frequency-magnitude relation of, 7, 10, 13

H

Half-space
　elastic, 125, 126
　homogeneous, 23, 81
　impermeable infinite, 78, 79
Hangingwall
　hard lava, 17
　soft lava, 15, 18
Hard-rock environment, 71
Haskell rupture model, 114
Hazard
　analysis of, 95
　assessment of, 94, 99
　predict, 69
　reduction of, 19
　rockburst, 17
　seismic, 15, 68
Heavy-tail distributions, 123
Herschel-Bulkley, 116
Heterogeneity, 63
　elastic, 126
　fault slip, 130, 136
　slip distribution, 111
　source, 108, 137

spatial, 114, 156
structural, 125
Hodograph transformation
canonical, 128
Homogeneous medium, 85
Horizon
oil-producing, 90
producing, 79, 83
Horizontal vector velocity
depth-averaged, 117
Hubbert-Rubey principle, 73
Human activities, 55, 70
Hybrid moment tensor inversion, 22
Hydration, 77
Hydraulic fracturing, 55
experiments on, 57, 74
flow conduits, 73
microseismicity, 91
permeability, 56
rock, 74
Hydraulic isolation, 90
Hydrocarbon production, 55, 75
Coalinga, 91
commercial, 67
correlation with, 62, 96
destructive earthquakes, 88
events induced by, 68
Gazli earthquakes, 56, 93
induced seismicity, 70, 93
midcrustal earthquakes, 57
seismic risk, 81
Strachan, 80
USA and Canada, 73
Valhall field, 71
Hydrodynamic response
coastal, 108
local, 152
Hydrodynamic theory
long-wave, 113
Hydrodynamics
complex, 108
Hydrophones, 71
Hypergeometric equation
confluent, 118
Hypocenters, 11, 23
Hypocentral distances, 14
Hypocentral separation, 6

I

Images
4D, 96
LANDSAT and SPOT, 92
seismic, 96

Imaging
4D seismic, 96
time-lapse seismic, 98
tomographic, 70
Immature fault zone, 93
Imogene, 61
heterogeneity, 63
Pennington model, 86
Impedance matching, 152
Incipient pillar, 25
Induced seismicity
patterns of, 55
Inelastic effects
presence of, 83
Initial phase
duration of, 41
search for, 40
Innovation Award, 71
Institutes
Central Mining, 6
Occupational Safety and Health, 5
Instrumentation, 96
high-quality, 55
low-noise, 71
Intact rock, 15
International Seismological Centre, 62
Inundation
multiple waves, 128
review of, 120
Inundation zone, 128
Inversion
fast and automatic, 44
moment tensor, 22
Richardson, 26
source time functions, 28, 29
Teyssoneyre, 27
Trifu and Shumila, 25
waveform, 23, 24
Isostasy, 60, 91
Isostatic imbalance, 91. *See also* isostasy
Isotropic component, 24, 26, 29
positive, 27, 28
source, 23
zero, 22
Isotropic medium, 114
Isotropic spectra, 152
ISS International, 2
seismic system of, 4

J

Japanese research group, 1

continuous monitoring, 3
investigation by, 40
observation of, 45
seismogenic experiments by, 2
Jogs, 20
Johannesburg
International Symposium at, 1

K

Kentucky, 71, 74
Kettleman North Dome, 57, 60, 91
Kidd Deep Copper-Zinc Mine, 25
Kingston
International Symposium at, 1
Klerksdorp
investigation, 4
Kopanang Gold Mine, 23
scattered mining, 19
Kloof Mines
cumulative frequency, 17, 18
Kolar gold fields, 2
Champion Reef Mine of, 42
study of, 5
Koyna dam, 55
Krakow
International Symposium at, 1
Kronecker delta, 114
Kummer's equation, 118. *See also* hypergeometric equation

L

Lacq, 58
comparison, 84
depth sections, 64, 65
fluid injection, 84
gas pressure plot, 84
heterogeneity, 63
induced seismicity, 56
location, 63
reverse faults, 87
Segall's model, 83
stiff reservoir, 81
Laguerre polynomials, 118
Lamé's first constant, 114
Laminar flow
equation for, 115
Landslide. *See also* turbidity current
basal friction, 127
debris avalanche, 116, 117
debris flow, 115, 128

Law
  decay, 152
  friction, 112, 157
  Green's, 134, 135
  power, 42, 116, 121, 133, 149
  runup, 111, 118
  slip-dependent constitutive, 35
  spectral decay, 122
Layer
  horizontal, 79, 89
  permeable, 78
  seismogenic, 91
Leaky modes, 118, 150, 157
Leaky shelf waves, 152
Legal implications, 97
Leveling, 83
Lévy, 123
Lévy α-stable, 122
Linear relationship
  fluid extraction, 79
  pore pressure, 81, 83, 84
Literature review, 2
Lithological units, 21
Localization
  event, 71
  improvement of, 66, 72
Location, 63
  active faults, 71
  aftershocks, 91
  brittle deformation, 87
  epicenters, 64, 66
  event, 71
  hypocenters, 75
  technique for, 72
Location algorithms, 72
Log-normal distribution, 133, 134, 157
Loma Prieta, 127
Long Valley Caldera, 34
Long-term correlations, 15
Lucky Friday Mine, 26

M

Magnitude, 55
  Alberta, 80
  destructive event, 91
  event, 68
  future earthquakes, 85
  low, 71
  moderate to large, 75, 94
  range of, 57
  small to moderate, 83
  subsidence, 97, 99
Main shock, 90

Main shocks, 42
Mass diffusivity, 79. *See also* mass flux
Mass flux
  constant, 79
Master event location methods, 6
Measurements
  BPR time-series, 140, 141
  deep-ocean, 157
  far-field, 144
  geodetic, 121
  runup, 133, 134
  seafloor deformation, 125, 127
  shore-based, 138
  tide gage, 128, 153
  wave tank, 111
Mechanics. *See also* Temporal
  fracture, 19
  high-resolution, 11
  micro, 15
  rock, 43
Mechanism, 28, 29, 75, 78. *See also* Pure shear
  caving, 19
  deviatoric, 12
  dip-slip, 27
  fault, 88
  focal, 10, 11, 20, 44, 80, 91, 92, 95
  hydrocarbon production, 90
  non-deviatoric, 23
  Pennington, 85
  solution of, 24
  source, 1, 3, 9, 20–27, 43, 65
  stress corrosion, 77
  stress diffusion, 42
  strike-slip, 26
MEQ-800, 94
Method, 13, 14
  analysis, 11, 44
  DD, 7, 14
  excavation, 43
  finite element, 19
  least-squares, 30
  Makowski, 32
  Mansurov, 41
  numerical and seismic, 18
  original, 22
  projected Landweber deconvolution, 30, 31, 33
  pseudo-spectral, 31, 33
  single-event location, 8
  standard absolute location, 6
  statistical, 3

Microseismicity, 55, 72
  correlation with, 66
  epicenters of, 72
  fluid injection, 75
  hydraulic fracturing, 91
  hydrocarbon fields, 57
  monitoring of, 56, 71, 96
Midcrustal Earthquakes
  controversy surrounding, 87–93
Midway Island
  cumulative distribution, 154
  density distribution, 154
  spectra, 155
  time series, 153
  wave heights at, 152
Mine design
  aim of, 18
  basic assumption for, 44
  influence of, 2
  numerical model, 43
Mining, 55
  human activities, 88
  industry of, 70
  laws of, 95
  map, 70
Mining operations, 15–21
Miningtek, 4
Minneapolis
  International Symposium at, 1
Mixture theory, 111, 115, 116
Modeling, 21
  force dipole, 26
  geomechanical, 96, 97, 99
  micromechanical, 24
  numerical, 1, 3, 17, 18, 24, 43, 44
  poroelastic, 64, 85, 89
  rock mass, 18, 23
  seismicity system, 42
  source, 93
Mohr diagram, 73
Mohr-Coulomb argument, 73, 74
Mohr-Coulomb model, 75
Moment tensor inversion, 23
  approach of, 25
  focal mechanisms, 24
  regional body waveforms, 27
  seismic events, 26
  technique of, 22, 44
Monitoring, 55
  acoustic emission systems, 11, 44
  continuous, 3, 8, 10

Index

4D, 96
Lorraine, 18
microseismic (MS), 14
passive, 71
petroleum reservoirs, 2
rock mass volumes, 5
seismic, 1, 12, 19, 22, 23, 43, 57, 69
stress, 41
systems for, 4
techniques for, 56
Monopole displacement, 147
Monte Carlo sampling, 6–8
Montebello, 60
   anticlines, 91
Monterey Canyon 1989, 127, 128
More Than Cloud (MTC), 72
MS and AE systems, 11
Multiplets
   seismic, 11

N

Narrow-banded waves, 157
NASA seismic array, 75
Near field
   broadside and oblique, 120–138
   devastating tsunamis, 111
   map view, 109
   single station records, 150
   source at, 151, 154
Nests, 10
Netherlands
   hydrocarbon fields, 68, 69
   Rotliegendes, 56, 58
Network
   eight-station telemetered analog, 75
   French National, 83
   local seismic, 57, 75
   Permian Basin, 68
   Uzbek, 92
Nodal planes, 26
   near vertical, 27
   right-lateral, 74
   steep, 87
   subvertical, 25, 27
Noise, 71, 96
Nominal coefficient
   friction, 112
Non-double-couple, 22
   focal, 27
Nonlinear effects, 82, 99
Normal strain, 3

North Ural Bauxite Mines, 42
Northparkes Endeavor, 19
Northridge, 123
Nova Scotia, 127
Nuclear explosions
   testing of, 3
   underground, 2
Numerical model
   calibrated, 43
   3D, 19
   damage-driven, 14
Numerical modeling
   approach of, 24
   design criteria, 18
   modern, 43
   rock mass behavior, 17, 44
   rock mass response, 1, 3

O

Ocean-bottom-seismometer cable, 71
Optimization, 32
Oscillations. See also water-column
   low-frequency elastic, 125
   small, 143
Outer rise, 111, 138
Overburden stress, 24

P

Passive monitoring
   applications of, 71
Patterns
   aftershock, 93
   fault, 71
   flow, 96, 98
   fracture, 56
   injection, 74, 75
   production, 57, 94
   seismicity, 57, 68, 72
   spatial, 62
   temporal, 64
PC based seismic monitoring, 5
Period
   dominant, 152
   interseismic, 112
   natural, 153
Periphery, 92
   fixed, 87
Permanent deformation, 113
Permeability, 79
   high fracture, 83
   hydrocarbon reservoir, 56

low, 90
   probability distributions for, 97
Permian basin
   induced seismicity, 56, 67
   oil and gas fields, 68
Perth
   International Symposium at, 1
Perturbation
   production-related, 56, 64
   stress, 80, 81, 83, 88
Petrographic analysis, 21
Petrographic evidence, 21
Petroleum Development Oman, 72
Phenomena
   man-made and natural, 2
   precursory, 1, 3
   reliable, 43, 45
   shock unloading, 21
Phenomenology (tsunamis)
   far-field regime, 138–156
   near-field regime, 120–138
Pillar. See also Incipient pillar
   barrier, 9
   carnallite, 27
   crown, 43
   dip, 19
   sill, 25
Planar fault, 110
Point-source inversion, 23
Poisson equation, 81
Poisson ratio, 80
Polarities
   body-wave, 22
   first, 23
Polarity, 134
Polkowice, 11, 30
Pore pressure
   conditions, 158
   Coulomb friction, 117
   distributions of, 157
   effects of, 112
Poroelastic equations, 80
Poroelastic model
   fluid extraction, 79
   Lacq, 83
   linear, 98
   Segall, 89, 90
Poroelasticity, 55
   theory of, 78
Porosity configuration, 78
Porous limestone
   offset of, 86

Post-failure dynamics, 115
Post-seismic earth responses, 121
Potash mine
  monitoring system, 5
  Teutschenthal, 27–29
  Upper Kama, 4
Precursory phenomena, 1
  large seismic events, 43, 45
  review of, 3
Prediction
  critical pressure, 74
  effects of fluid extraction, 81
  induced seismicity, 83
  model, 79, 84, 85
Pre-existing discontinuities, 98
  lithological, 84
Pressure, 74, 84
  critical, 74
  drop, 69, 84
  fluid, 73, 75, 86
  pore, 56, 57, 74, 80–83, 85, 89, 90, 94
  pore fluid, 15
  subvertical, 25
Pressure drop, 69
  oil and gas fields, 86
  plot of, 84
Pressure gages
  deep-ocean bottom, 108
Probability distributions, 84
Projected Landweber
  deconvolution, 30
  source duration, 31
  source time functions, 33
  time-domain approach, 30
Projects
  4D seismic, 96
  carbon sequestration, 55
  microseismic monitoring, 71
  More Than Cloud (MTC), 72
  secondary recovery, 90
Pseudo-spectral method, 31, 33
Pseudo-velocity, 14
Puerto Sandino, 129
Puget Sound, 126
Pure shear mechanism, 22
Pyrenees, 83, 85

Q

Quartz, 77
Quasi-harmonious oscillations, 76
Quaternary, 93

R

Radiated energy
  dynamic measure, 32
  earthquakes, 35
  estimation of, 5, 44
  ratio of, 34, 39
  short-term precursor, 42
Rajendra Colliery, 5
Rangely, 60
  experiment at, 62, 73, 74
  solutions for, 74
  spatial clustering, 63
RaSiM, 1, 43
  review of, 2
Ray path
  least-time, 140
Ray theory, 134
Rayleigh waves, 27
  oceanic, 113
Region
  focal, 90
  Gazli, 91–93, 95
  seismically active, 92
  seismically inactive, 93
Regional occurrence, 67–69
  induced seismicity, 69
Regression
  least-squares linear, 35
  two-stage, 14
Relations, 15
  energy-moment, 35
  ground-motion prediction, 14
  Gutenberg-Richter, 7, 10, 13
  magnitude-frequency, 5, 18
  scaling, 1, 3
  statistical requirements for, 19
  time-space-strength, 43
Relative moment tensor inversion, 22
Relative Source Time Function (RSTF), 29
Relaxation
  logarithm of, 41
  monotonic, 40
  spatial, 42
Reservoir
  axisymmetric, 80, 81, 84
  depletion of, 56, 87
  impoundment, 55
  infinitely extended horizontal, 78–80
  management, 77
  modeling, 80, 86, 87, 95

properties of, 71, 81, 96, 97, 99
  rocks, 69, 77, 78
  subsidence, 67
Reservoir compartmentalization, 96
Reservoir properties
  comparison of, 94
Reservoirs, 2
Resonance, 107. *See also* shelf resonance
  Crescent City harbor, 153
  harbor, 154
  scattering and, 118
  trapped edge waves, 150
  trapped modes, 157
Response
  pseudo-velocity, 14
  recording instrument, 28, 44
  rock mass, 1–3, 11, 18, 41, 43
  rock mass impulse, 32
  short-term seismic, 15
  temporal, 21
Rheology. *See also* bilinear
  Bingham plastic, 116
  granular, 111
  modulus, 112, 121, 130
  Newtonian, 117
  rigidity, 125
  viscoplastic, 115
Rigid block model, 111
Rock mass response
  excavation stresses, 44
  microseismic sequences, 11
  mining, 1, 18, 43
  monitoring of, 41
  numerical modeling of, 3
  production, 19
  stress and strain changes, 2
Rock-flour, 21
  cataclastic, 20
Rockburst, 4
  compact, 20
  damaging, 5
  excavations resistant to, 17
  geometry of, 21
  hazard of, 17
  in situ studies of, 44
  International Symposium on, 1
  Lucky Friday Mine, 26
  magnitude of, 43
  research on, 2
  rock mass prone to, 19
Romashkino, 59

Index

correlation, 75
epicenter map, 76
seismic activity at, 76
spatial clustering, 63
superposition, 77
Roof strata
boreholes, 41, 45
fracture propagation in, 5
Root-mean-square
water-level elevation, 141
wave height, 142
Roswinkel, 58
active field, 68
regional occurrence, 69
Rotenburg, 58
relocalization, 72
Rotliegendes, 58. *See also*
Netherlands
faulting, 64
geometry, 87
induced seismicity, 56, 67
surface subsidence, 96
Roughness
fault, 121
seafloor, 149
slip distribution, 122
Rudna mine
apparent stress, 36
blind deconvolution, 32
events at, 10, 29, 31, 32
investigation at, 40
multiplets, 11
networks, 30
source time function, 33
static stress drop, 37
Rupture pulses
self-healing, 113
Russia
seismic monitoring in, 1, 3, 43
Upper Kama potash deposits, 4, 27

S

San Andreas, 89
San Joaquin, 90
Sand cover
thickness of, 87
Sand-silicone box, 87
Satellite altimetry, 158
Savage-Wood, 39
Scaling relations, 1
apparent stress, 36
discussion of, 3
seismic events, 35

Scarps
fault, 113, 125
head, 115
Scattering coefficient, 149
Scattering index, 149
Seafloor displacement
creeping, 110
impulsive, 109
Seattle, 125, 126
Secondary recovery
fluid injection, 56, 73
measures of, 76
oil, 75
projects for, 90
Sedimentary basins, 56, 67
Sedimentary environment, 71
Segall's cross sections
original and modified, 88
Segall's models, 78–84
diffusion equation, 79
poroelastic model, 79
Seismic array
downhole geophone, 72
NASA, 75
onshore and offshore, 95
permanent, 71
Seismic catalogs
Monte Carlo simulated, 13
synthetic, 18
Seismic efficiency
conventional, 39
Seismic events
energy classification of, 76
Seismic instruments
broadband, 120
Seismic management, 17
Seismic monitoring
continuous, 8
expansion of, 43
mines, 1, 12
new techniques in, 3
origin of, 4
rock mass damage, 14
stress-induced microcracking, 11
systems for, 19
underground systems, 5
University of Utah, 10
use of, 22
Seismic network
local, 83
map of, 94
Tatneftegeophysica, 75
vertical seismometers, 74

Wetmiller, 80
Seismic nucleation. *See also* Slow
initial phases
phases of, 40
Seismic signals, 32
Seismicity
accelerated, 42
average upper size of, 9
correlation with production, 57, 62
induced, 21, 22, 55–99
International Symposium on, 1
level of, 2
micro, 6, 15
mines, 43
mining factors affecting, 3
mining-induced, 41
modes of, 7
monitoring, 23, 29
natural, 4, 57, 68, 70
observed, 11
parameters of, 19
pattern, 56, 57, 68, 72, 75
pattern of, 8
potash mines, 5
recording and analysis of, 17, 44
study of, 35
two types of, 10
Seismograms, 40, 127
digital, 34
inversion of, 108
observed, 125
STF from, 27, 44
Seismographs. *See also* MEQ-800
CEIS Lithoscope, 94
digital and conventional, 80
portable, 75
Seismometers
borehole, 69
vertical, 74
Sensors
continuous recording from, 5
locations of, 25
seismogenic areas, 2, 40
triaxial, 26
underground seismic
networks, 29
uniaxial, 23
vertical, 30
Sequences
aftershock, 12, 13
microseismic, 11
Seventy-six oil field, 71

Shallow-angle faults, 120
Shapiro-Wilk
  test of, 142, 152
Shear modulus, 80
Shear wave splitting
  analysis of, 41
  observation of, 45
Shear waves
  anisotropy of, 65
  speed of, 67
Shelf resonance, 154
Shoaling amplification, 134, 156
Shock unloading, 21
Shuaiba, 59
  epicenters at, 66
  faults of, 72
  heterogeneity, 64
Signal-to-noise ratio, 96
Signals. *See also* Seismic signals
  amplitude of, 40
  log-periodic, 42
  quality of, 5
  short-term, 32
Silicate rocks, 77
Silicon-oxygen, 77
SIMRAC, 2
Simulations
  complex hydrodynamics, 108
  3D, 125
  earthquake slip dynamics, 113
  high-order numerical, 111
  oceanic Rayleigh waves, 113
  stationary wave sequence, 142
  tsunami coda, 157
Single-source inversion, 22
Site response, 134
  local, 152, 154
  nearshore, 157
  specific, 156
Skempton's pore pressure
  coefficient, 80
Sleepy Hollow, 60
  hypocenters, 75
  swarms at, 63
  time lapse, 64
Slip
  barrier to, 85
Slip-weakening distance
  critical, 112
Slow initial phases, 40
Slow tsunami earthquakes, 112, 138
  behavior of, 121
  occurrence of, 113

Smoothing procedure
  space and time, 42
Solutions
  analytic, 78, 80
  fault plane, 74, 93
  focal plane, 74
Source
  pore pressure, 82
  shallow, 81
  toroidal and point, 83
Source mechanism, 1, 22–27
  discussion of, 3
  seismic, 44
  three main types of, 43
Source parameters, 35, 36
  calculation of, 4
  discussion of, 3
  estimated, 20
  estimation of, 30
  measurement of, 13
  near, 13
  seismic event, 44
Source signal, 32
Source time function (STF), 1
  calculation of, 33
  decomposition of, 27–32
  discussion of, 3
  extraction of, 44
  fault length, 37
  nonlinear determination of, 23
South Africa
  Bambanani Mine, 6
  earthquake-generation
    process, 40
  East Driefontein Mine, 16, 18
  East Rand Propriety Mines, 20
  Far West Rand, 26
  gold mines in, 2, 38
  in situ studies, 44
  International Symposium at, 1
  investigation committee, 4
  Japanese research group, 45
  Kloof Mines, 17
  Kopanang Gold Mine, 23
  Mponeng Gold Mine, 15, 36, 37
  Oryx Gold Mine, 22
  seismic events, 35
  tests, 42
  Ventersdorp Contact Reef, 19
  Welkom Gold Field, 21
  Western Deep Levels, 41
  Witwatersrand Basin in, 11, 17
South African gold mines, 1

apparent stress, 36
data sets from, 36
deep gold mine, 42
earthquake-generation
  processes in, 40
new results on, 3
post-seismic deformation, 45
slope coefficient in, 35
small seismic events in, 35
static stress drop, 37
2000-3600 m depth, 2
Space-borne technology, 158
Space-time
  mining activity in, 8
  patterns of, 42
Space-time correlation, 42
Space-time-magnitude
  distributions of, 8
Spatial patterns, 62
  clustering, 68, 83, 85, 91
State
  Coulomb stress, 85
  in situ stress, 74
  natural stress, 83
  undrained, 80
Static fatigue, 77
Statistical Distributions. *See also*
  Lévy
  Cauchy, 123
  exponential, 136, 140, 141, 154, 157
  fractal, 122
  Gaussian (normal), 142, 152, 154
  power-law, 133
  Rayleigh, 142–144, 154
  von Kármán, 122, 149
Step bathymetry model, 136
Stiff reservoir, 81
Stiffness, 97
Stiffness ratio, 69
Stochastic slip
  distributions of, 123
  models of, 108, 121, 122, 157
Stokes mode, 118
Stopes
  active, 41
  area of, 21
  close vicinity of, 2, 40
  coseismic, 24
  deep-level, 15
  faces of, 5, 7
  longwall, 20
  margins of, 27

Index

planning and sequencing of, 19
rock mass around, 42
Strachan gas field, 58
    hydrocarbon production, 80
    reverse faults, 87
    temporal patterns, 64
Strain, 66, 85. *See also* normal strain
    axial, 22
    changes in, 43, 44
    energy of, 17
    maximum principal, 40
    plane, 89
    shear, 3, 116
Strainmeter, 40
Strata
    oil-bearing, 90
Strategies, 73
    secondary recovery, 73
Stress
    accumulation of, 85
    bending, 111
    Coulomb, 83, 85
    deviatoric, 89
    effective, 73, 74, 78
    induced, 85
    mean, 79, 89
    natural, 81
    Newtonian viscous fluid, 117
    perturbation in, 56, 64, 80, 88
    pre-existing, 64, 68
    shear, 112, 113
    tectonic, 83, 94
    yield, 116, 127
Stress component, 83
    principal, 85
Stress corrosion, 77, 98
Stress diffusion
    concepts of, 1, 3, 45
    mechanism of, 42
Stress loading, 75
Stress-transfer effects, 111
Strike-normal direction, 124
Subaerial flows, 115
Subduction zone, 127
    continental, 148
    Hokkaido-Kuril, 152
    inter-plate thrust of, 123
    tsunamigenic earthquakes, 111, 120
Subevents
    factorization of, 28, 29
Submarine environment, 115–117

Submarine mass movements, 112
Subsidence, 57, 60, 65, 87, 93, 96
    amount of, 81
    cessation of, 87
    down-dip, 124, 125
    localized areas of, 138
    magnitude of, 97, 99
    massive, 86
    reservoir, 67
    seabed, 66, 67
    surface, 79, 84, 87, 89, 90, 98
    theoretical and experimental, 83
    vertical, 77, 78
Subsidiary faults, 112. *See also* inter-plate thrust
Superposition
    seismic activity cycles, 85
Swarms, 10
Symposium
    Joint Japan-Poland, 2
    RaSiM, 43
Systems. *See also* ISS
    acoustic emission, 11
    data acquisition, 3
    dynamic range seismic, 40
    elementary, 32
    modern digital, 44
    recording, 41
    seismic monitoring, 43
    seismicity, 42
    smoothing of, 15
    support, 17

T

Tabular sequential grid, 15
Tatneftegeophysica seismic service, 75
Techniques, 18, 42
    adaptation of, 3
    automatic event location, 6
    deconvolution, 28
    EGF, 29
    empirical Green's function, 35
    fast and automatic inversion, 44
    group location, 6
    moment tensor inversion, 22
    new, 1, 3, 43
    nonlinear, 30
    room-and-pillar, 20
    spectral ratio, 34
    time-domain, 31
Tectonic activity, 68

Tectonic compression, 91
Teleseismic *P* waves, 120
Temporal fracture, 11
Temporal patterns, 62, 64
Theory
    analytic, 108, 130, 150
    constitutive, 117
    dislocation, 123
    elastic displacement, 144
    long-wave hydrodynamic, 113
    ray, 134
    rheology and mixture, 111, 115, 116
    seismic coda, 141
    solitary wave, 120
Theory of poroelasticity, 78
Thrust, 91
Tidal stresses, 2
Tide gage
    coastal, 108
    Corinto, 129
    Crescent City, 136, 150, 151
    data of, 138
    locations of, 154
    measurements of, 128, 153
    North Pacific map, 140
    records of, 127, 129–131, 152, 153, 155
Tilting, 145
Time lag, 67, 80
Time lapse in onset, 64, 75
Time-dependent data, 96. *See also* time-lapse
Time-space-strength, 43
Tomographic imaging, 70
Trail Mountain, 8
    coal mining in, 14
    displacement waveforms, 24
    epicenter map, 9
    mining seismicity at, 10
Transmissivity, 149
Transoceanic tsunami, 108, 156
    coda for, 148
    indicator of, 112
    map view, 109
Trans-Pacific tsunamis, 155
Trapped waves
    absence of, 152
    coastal wave, 118
    multiple, 156
    non-, 128
    resonance of, 150, 157
Traps
    structural, 86

Trench-normal convergence, 112
Triad
  nonlinear coupling of, 154
Trough. *See also* wave height
  amplitude of, 143, 153
  wave, 156
Tsunami. *See also* far field
  Airy function, 144, 146
  Airy phase, 136
  amplitude, 121, 141
  bottom friction, 120
  coastal (nearshore) response, 107, 157
  coda, 138, 140–143, 154, 155
  Crescent City (California), 150, 151
  decay (attenuation), 118, 149
  dispersion, 144, 148
  earthquake generation (seismogenic), 110, 158
  edge waves, 119, 154
  e-folding, 141, 152
  envelope, 141, 142
  first (direct) arrival, 129, 131
  focusing, 138, 149
  Green's law, 134, 135
  group velocity (speed), 113
  hazard, 156
  initial conditions, 148
  inversion (slip), 125
  landslide generation, 115–117
  nonlinearity, 108, 118
  non-trapped, 128, 150
  oblique, 112, 136–138
  phase velocity (speed), 109, 110, 156
  radiation, 145, 147, 150
  ray path, 140
  reflections, 127, 148
  resonance, 153, 154
  runup, 111, 113, 128, 132, 133
  scattering, 130, 144
  shallow-water wave equation, 117
  shelf modes (waves), 118, 152
  spectrum, 118, 122, 123, 137, 139–143, 155
  surveys, 130, 133
  turbulence, 120
  water-level (wave) elevation, 109, 140, 141, 143
  wavefield, 112, 125, 134, 140
Tsunami deposits, 158
Tsunami earthquake
  definition of, 107
  fault mechanics of, 158
  generation of, 111–114, 156
  observations of, 128–138
  radiation pattern of, 147
  source observations, 120–127
Tsunami problem
  schematic view of, 114
Tsunami shadows
  detection of, 158
Tsunamis. *See also* Monterey Canyon
  Cape Mendocino 1992, 136, 151
  Chile 1960, 150, 155
  Gorda Plate 2005, 152
  Grand Banks 1929, 127, 128
  Hokkaido 1993, 107
  Kamchatka 1952, 136
  Kuril 2006, 140, 148, 150, 151, 153
  Kuril 2007, 138, 140, 141
  Nicaragua 1992, 129, 132, 133
  Nihonkai-Chubu (Sea of Japan) 1983, 136
  Sulawesi 1996, 133
  Sumatra (Nias) 2005, 135
  Sumatra-Andaman (Indian Ocean) 2004, 108, 112, 114, 145, 146, 154
  Tokachi-Oki 1952, 133, 135, 156, 157
  Tokachi-Oki 2003, 125, 130, 131, 158
Turbidity current, 115, 128
Two-step inversion, 23

U

Uinta Basin Observatory, 74
Uncertainty
  aleatory, 157
  epistemic, 158
Under-sampling
  temporal, 112
Underground Research Laboratory
  blast excavation, 6
  common rock volume, 11
  data from, 15
  data sets from, 35
  seismic events from, 18, 19, 22
  study by, 21
Units
  geological, 56
  shale, 90
University of Grenoble, 93, 94
University of Utah
  seismic network of, 8, 10, 14
Upheavals
  floor, 27
Uplift
  continuous, 93
  localized areas of, 138
  recent, 91
  region of, 130
  ring of vertical, 81
  up-dip, 124, 125
  vertical, 124
Upper Silesian Coal Basin
  a posteriori PDF, 7, 8
  borehole deformations in, 41
  earthquakes in, 10
  events in, 6
  longwall coal mining in, 27
  seismicity of, 9
  Wujek and Ziemowit, 11
U.S. Geological Survey, 75, 90

V

Valhall, 59
  anisotropy, 65
  chalk formation, 71
  heterogeneity, 63
  instrumentation, 96
  seabed subsidence, 67
  spatial variations, 67
Vegetation discoloration, 128
Viscoplastic fluid, 115
Volcanic-dominated islands, 117
Volcanogenic landslides, 127
Volume
  cone-shaped, 87
  fluid, 75
  fractured, 70
  pore, 78

W

War-Wink, 61
  location, 63
  seismicity in, 68
Waste injection, 55
Water flooding, 78
Water injection, 75
Water-column oscillations, 125
Wave height, 113
  crest-to-trough, 142

cumulative distribution of, 144, 154
deep-ocean tsunami, 158
empirical, 152
tsunami coda, 143
Wave profile, 130. *See also* strike-normal
cross-shore, 118, 119
leading, 110
tsunami, 134
Waveforms
characteristics of, 150
complex, 11
deep-ocean, 140
far-field, 148, 158
initial, 138
initial phases, 40
inversion of, 24, 157
microseismic, 42
observed, 31
offshore, 130
processing of, 44
recording of, 5
regional body, 27
seismic, 127, 156

shape of, 146
solitary, 120
spectral analysis of, 34
synthetic, 33
three-component, 23
tsunami, 108, 121, 144
Wavemaker, 111
Waves. *See also* shear
compressional, 67
elastic, 71
Welkom, 42
Bambanani mine, 40
Matjhabeng mine, 21
Western Deep Levels, 20, 41
Whittier Narrows, 91
Wiener-Khintchine, 122
Wilmington, 60
hydrocarbon production, 94
new faults, 65
surface subsidence, 77, 82, 96, 97
temporal patterns, 64
Witwatersrand Basin
East Rand Propriety Mines, 20
mining areas, 11

scattered mining, 19
ultradepth, 17
Ventersdorp Contact Reef, 19

Y

Yield strength
finite, 117

Z

Zone. *See also* Gouge zone
boundary, 91
complex, 72
deformation, 20
dipping, 95
extensional and compressive, 79
fault, 21, 34, 35, 90
fracture, 24
fractured, 90
high stress, 5
seismogenic, 10